Giorgio Piola

Formula 1

technical analysis 2008 2009

Even the best Hollywood scriptwriter would have had a hard time plotting the 2008 F1 season with its heart-stopping finale that saw Felipe Massa's hands on the championship crown for a few seconds before it was torn away from him as Lewis Hamilton passed Timo Glock at the last corner to take the 5th place that guaranteed him the world title. A stand-up fight between two drivers who would both have been worthy champions. The first coloured driver to win the Formula 1 World Championship gained a measure of compensation for all the goings-on that had blighted the 2007 season, while the Brazilian could take some satisfaction from an error-free season despite being penalised by the Rosse's inferior reliability compared with the Silver Arrows. To conclude this review of the sporting side of the season which we generally gloss over in this analysis, mention has to be made of the record number of race-winning drivers, no less than seven: Massa, Raikkonen, Hamilton, Alonso, Kubica and Vettel. The polish driver was actually in the running for the title for almost half the season before it became a private affair between Ferrari and McLaren. There were also a number of incidents never previously seen in a Formula 1 race that were more concerned with the

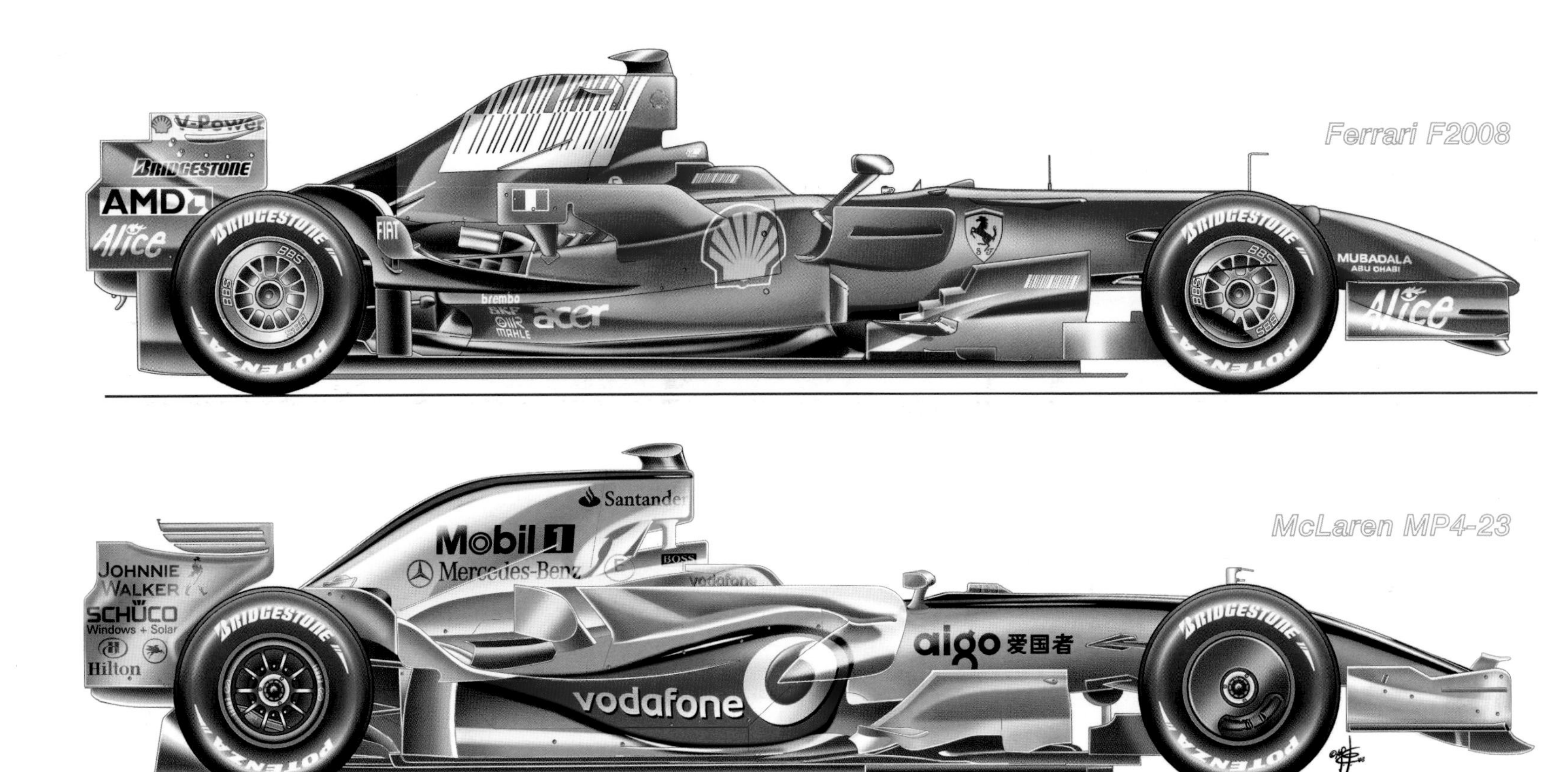

The 2008 SEASON

human rather than technical side of the sport such as the pit lane pile-up in Canada involving Raikkonen, Hamilton and Rosberg and then Massa's Ferrari dragging its refuelling hose with the Ferrari mechanics chasing behind in the Singapore Grand Prix. There were neither disputes nor scandals and the season featured technical innovations at every championship round. In technical terms, the championship was even more fascinating than the one before which had itself been particularly interesting thanks above all to the technical stability that (MES and the abolition of certain electronic aids apart) had permitted the teams to develop new features. The electronics ban favoured the introduction of new driver control systems such as the six levers behind the McLaren steering wheels (imitated by Renault and Honda) and the starting systems developed by Ferrari (see Cockpits chapter). The restrictions on electronics made driving styles more important and brought back drifting and on-the-limit handling, consequently making the racing more spectacular.

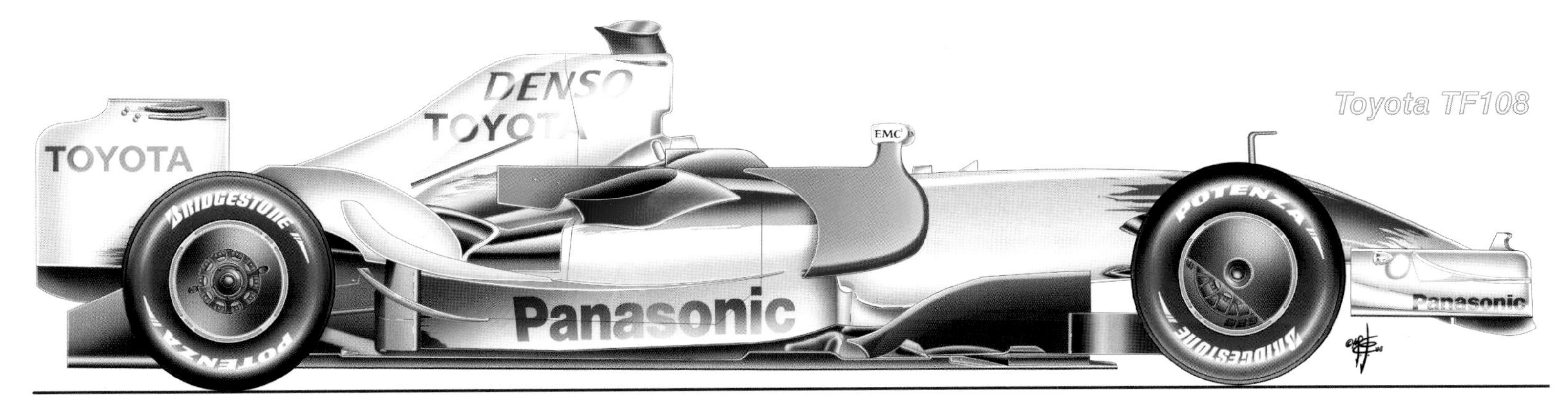

RESTRICTED AERODYNAMICS FOR 2009

In the final season of hyper-developed Formula 1 aerodynamics before the introduction of the severe restrictions announced by the Federation for the 2009 season, the teams outdid themselves in their wind tunnel research. Never have we seen as many new aerodynamic features as in 2008. The Ferrari-McLaren duel that was eventually won by the Rosse saw multiple micro-developments. In effect, at every race the two great rivals introduced mini aerodynamic packages to better adapt their cars to the various circuits. The most interesting aspect was that many of the new features introduced and subsequently copied originated not with the two leading teams – apart from the wheel covers introduced by Ferrari in 2007 and adopted across the board in 2008 – but with BMW, Red Bull and even the relatively uncompetitive Honda.

THE TORO ROSSO SURPRISE

The surprise of the season was undoubtedly Toro Rosso scoring its maiden Formula 1 victory at Monza with a fantastic performance from the young Vettel. Above all, the team took the liberty of finishing ahead of its progenitor Red Bull in the Constructors' Championship. This was largely due to solid technical direction based on the experience of the former Berger-era McLaren track engineer and ex-Ferrari man Giorgio Ascanelli; merit was also due to the Ferrari engine that out-classed its Renault rival on the Red Bull.

NEW SPORTING REGULATIONS

Confirmation of the requirement to use both kinds of tyre supplied by Bridgestone during the course of the race without the use of electronic aids made race tactics even more interesting and in this field Ferrari frequently enjoyed an advantage over its direct rival McLaren. The qualifying system was revised; the most important change concerning the abolition of fuel top-ups during the Q3 session, which was shortened to just 10 minutes and became spectacularly decisive in establishing grid positions with the fuel load with which the GP was to be started.

ABOLITION OF THE T-CAR

The T-car or spare disappeared from the pits in the 2008 season, although teams did bring with them two knocked-down examples to be assembled in the case of an accident. However, should a driver damage his car, the replacement would only be ready for the following day's session. The spare chassis was no longer taken to the circuit preassembled, but was stripped of all suspension components and so on.

SECOND HALF OF THE SEASON SACRIFICED

Many of the teams that found themselves excluded from the fight for victory deliberately sacrificed the second half of the season, devoting all their resources to the development of concepts for the 2009 car. This group consisted of Honda, Toyota, Red Bull and Williams. Renault's development of its 2008 car started late and it was always struggling to recover, although it did manage to save face with a single victory in the Singapore GP.

SPYKER-FORCE INDIA

In the latest name change for the team that was once Jordan and then Midland, Spyker became Force India in 2008 and was led by Mike Gascoyne through to mid season and as in 2007 was powered by Ferrari like the Toro Rosso team.

THE DEMISE OF SUPER AGURI

The last to join the GP circus, Super Aguri disappeared after four races. Looked upon with a degree of scepticism in 2006 as to all intents and purposes it was racing with a four-year-old chassis, it produced a minor miracle the following season in which the tiny Japanese outfit managed to upstage its mother-team, Honda. Lack of funds then put paid to the team's Formula 1 adventure immediately after the Spanish GP.

Honda RA108

Force India VJM01

Super Aguri 2008

In the fifth season after the parc fermé regimen was introduced between qualifying and the race, the use of the T-car and the spare chassis was restricted to those cases in which a driver had irreparably damaged his chassis on the track.
Moreover, the 2008 season also saw the introduction of the rule stating that in the case of the replacement of the chassis, the driver would not be permitted to take part in the successive practice session on the same day. For this reason, the spare chassis had to be completely stripped rather than being pre-assembled, as was the case in previous years to facilitate the replacement operation.
We have therefore reduced the chassis tables, as they would have frequently been very flat. There was confirmation of the tendency to build fewer chassis, the most coming from Ferrari with a total of 8 units.

FACTS AND FIGURES

CHASSIS BUILT

The tally of chassis built in the 2008 season saw Ferrari out in front with 8, but it should be noted that the last of these was constructed when the season was all but over as a Kers test-bench. McLaren and BMW followed with 7 units each. Then came Honda with 6, but here too, one of the chassis only took to the track at the end of the year with the Kers fitted. Toyota built 5 units and Renault, Red Bull and Toro Rosso 4 while Super Aguri made do with just 3. McLaren in the last 2 races took 3 spare tubs.

VICTORIOUS CHASSIS

There was a record number of winning teams in 2008: Ferrari, McLaren, BMW, Toro Rosso and Renault. The laurels for the most successful chassis were shared between two Ferrari units driven by Massa, both with 3 wins (267-269) and McLaren chassis (04) driven by Hamilton, then with 2 wins McLaren (06) with Hamilton. Then with a single victory each came Raikkonen fs Ferrari chassis (268-270), Hamilton fs and Kovalainen fs McLarens (02 and 05 respectively), Kubika fs BMW 03, Vettel fs Toro Rosso 03 and Alonso fs Renault 02.

Chassis F2008	265	266	267	268	269	270	271	272
First run	07-01-2008	20-02-2008	14-01-2008	25-02-2008	14-05-2008	14-05-2008	26-08-2008	17-11-2008
Km completed GP	0	0	5.983	2.105,6	6.826,3	8.616,8	2.079,8	2.079,8
Km completed test	9.177,3	4.564,4	4.404,4	4.000,2	1.701,2	2.515,5	2.547,3	3.522,9

FERRARI • F2008 • N° 1-2

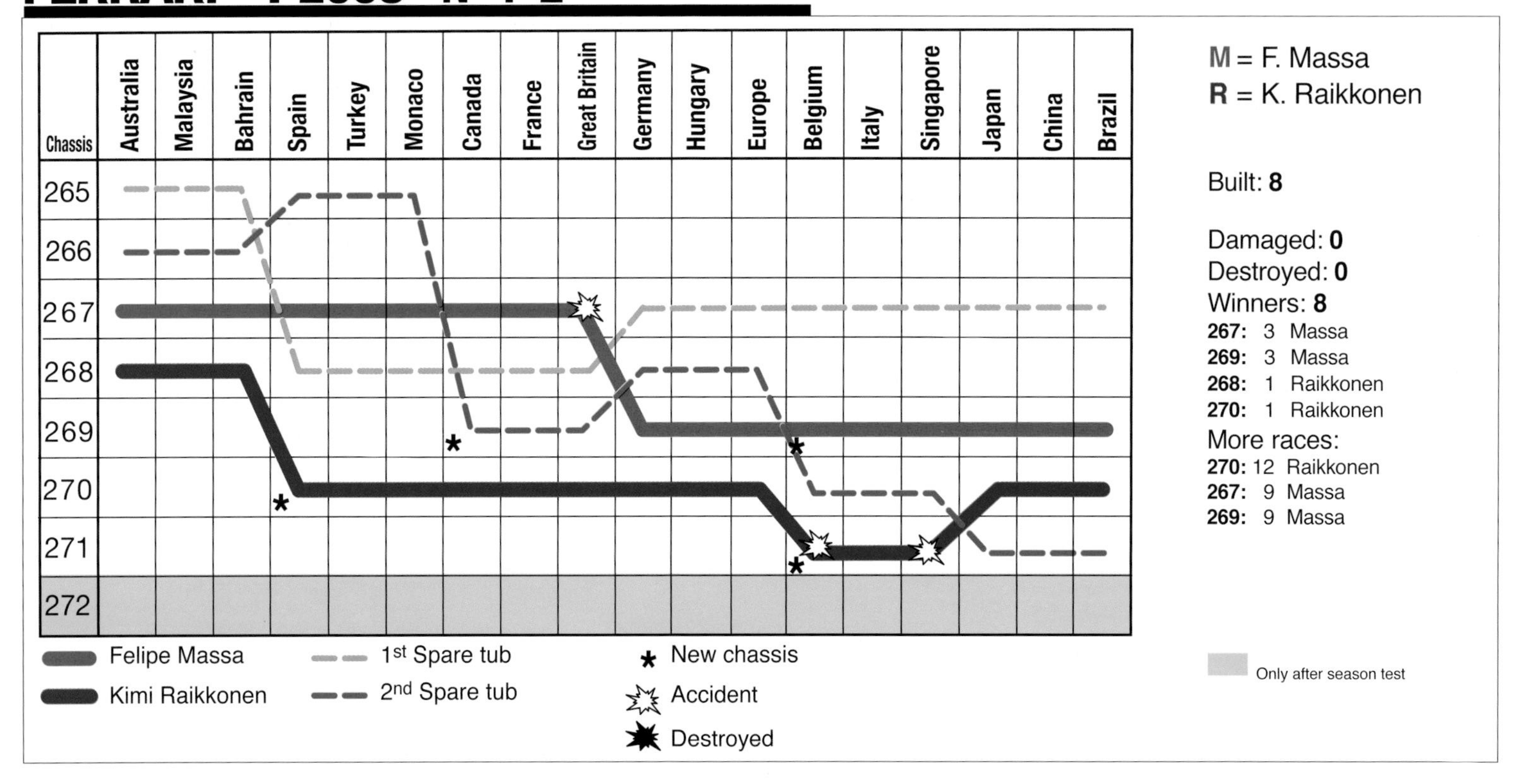

WHEELBASES

The car with the longest wheelbase was the Ferrari F 2007 at 3,295 mm, followed by the BMW at 3,235 mm and the Honda at 3,192 mm. Then came the McLaren at 3,187 mm, the Renault at 3,165 mm, the Williams at 3,159 mm, the Red Bull and the Toro Rosso at 3,140 mm and finally the Toyota with 3,134 mm.

	laps completed (%)	finishes	technical failures	accidents	test km	days
BMW	2196 (98,3%)	34	**0**	2	23079	48
Ferrari	2099 (93,9%)	28	**4** engine	4	23965	54
Williams	2091 (93,6%)	31	**1** engine	3	23673	46
McLaren	2078 (93,0%)	31	**4** eng.(2) - gearbox(1) - wheels(1)	1	23964	45
Toyota	1960 (87,7%)	29	**2** battery (1) - hydraulics (1)	5	22776	44
Honda	1921 (86,0%)	27	**3** disqualif. - gearbox - electr.	5	21264	43
Red Bull	1810 (81,0%)	28	**2** gearbox(1) - radiator (1)	6	20256	45
Renault	1778 (79,5%)	24	**4** engine - electron. - gearbox - brakes	8	22394	46
Toro Rosso	1697 (76,0%)	25	**3** exhaust - p. susp. - electr.	8	18400	37
Force India	1487 (66,6%)	18	**8** gearbox (4) - hydraulics (2)	10	14329	33
Super Aguri	326 (68,8%)*	5	**2** driveshaft - radiator	1	1346	4

* Only four races.

The most reliable team was BMW, completing an incredible record of 87.3% of the total number of laps scheduled over the course of the entire World Championship, with no DNFs due to mechanical failures. In second place came Ferrari, with 93.9% despite 4 engine failures. Third was Williams, despite a nondescript season, with 93.6%. McLaren was only fourth with 93%, but in the confrontation with Ferrari it suffered just 2 engine failures. The least reliable team was Force India with just 66.6% of laps completed.
The restriction to just 30,000 km of testing saw the teams significantly reduce the number of private test sessions.
In terms of number of days Ferrari led with 54 followed by BMW (48), Renault and Williams (46), McLaren and Red Bull (45), Toyota (44) and Honda (43). Way behind came Force India with just 18 days f testing.
The Safety Car was used on no less than 13 occasions, in 9 different races, for a total of 228.7 km.
Ferrari led the field in terms of kilometres covered with 23,965, closely followed by McLaren (23,964 km), Williams (23,673), BMW (23,079 km), Toyota (22,776 km) and Renault (22,394 km).

McLAREN • MP4-23 • N° 22-23

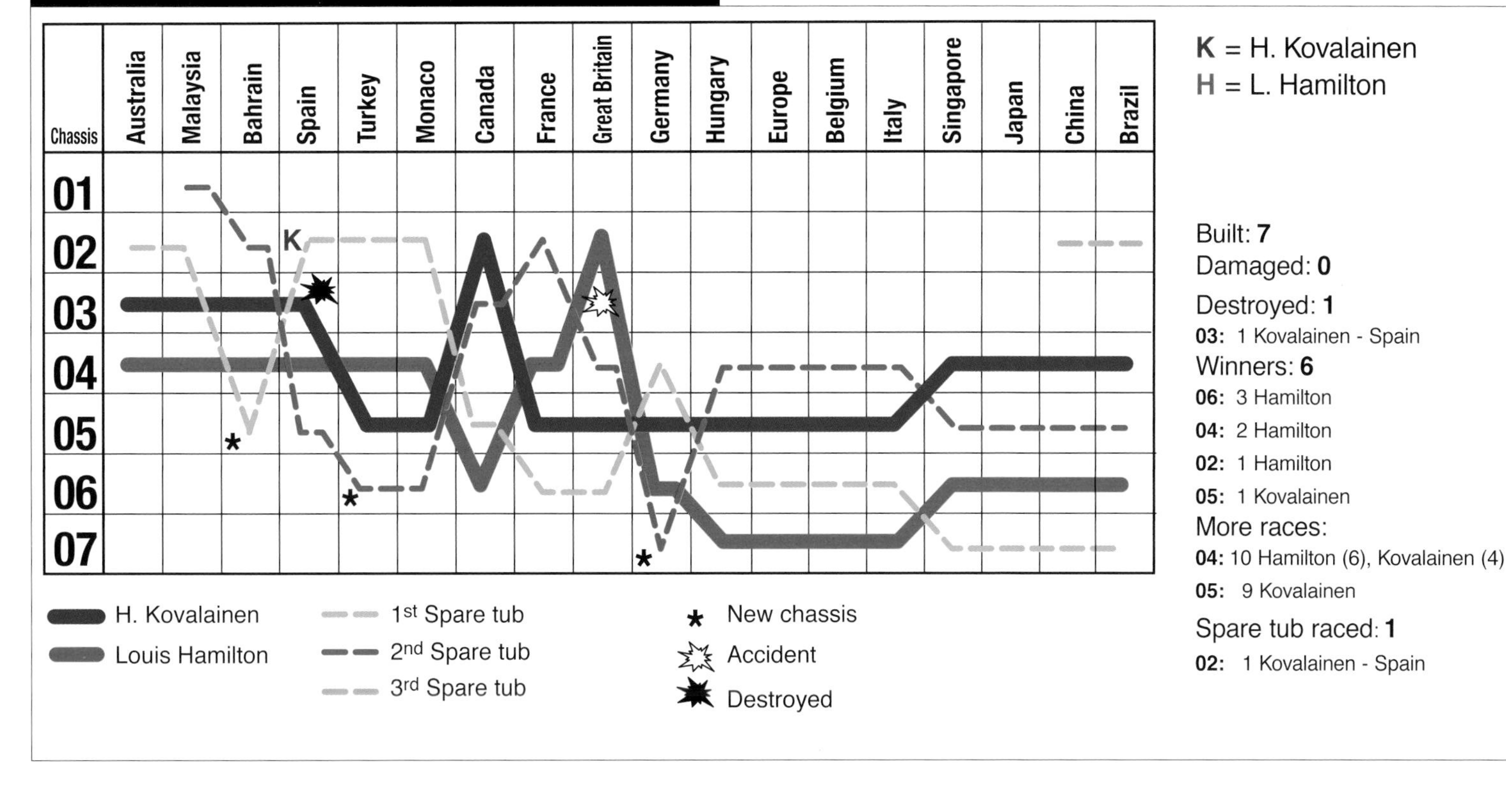

		1 - 2 FERRARI	22 - 23 McLAREN	3 - 4 BMW	5 - 6 RENAULT	11 - 12TOYOTA
CAR		F2008	MAP4/23	F1.08	R28	TF108
	Designers	Aldo Costa - Nikolas Tombazis - Gilles Simon	Jonathan Neale - Neil Oatley Paddy Lowe	W. Rampf M.Duesmann - W. Riedl	Pat Symonds - Bob Bell Rob White - Dino Toso	Pascal Vasselon Luca Marmorini
	Race engineers	Rob Smedley (1) Chris Dyer (2)	Phil Prew (22) Mark Slade (23)	G. Dall Ara (3) Antonio Cuquerella(4)	David Greenwood (5) Phil Charles (6)	Gianluca Pisanello (11) Francesco Nenci (12)
	Chief mechanic	Francesco Ugozzoni	Pete Vale	Urs Kuratle	Gavin Hudson	Gerard LeCoq
CHASSIS	Wheelbase	3195 mm*	3187 mm*	3214 mm*	3165 mm*	3134 mm
	Front track	1470 mm	1470 mm*	1460 mm	1450 mm	1425 mm
	Rear track	1405 mm	1405 mm*	1400 mm	1420 mm	1411 mm
	Front suspension	2+1 dampers and torsion bars	2+1 dampers and torsion bars	2+1 dampers and torsion bars	2+1 dampers and torsion bars	2+1 dampers and torsion bars
	Rear suspension	2+1 dampers and torsion bars	2+1 dampers and torsion bars	2+1 dampers and torsion bars	2+1 dampers and torsion bars	2+1 dampers and torsion bars
	Dampers	Sachs	McLaren	Sachs	Penske	Sachs - Toyota
	Brakes calipers	Brembo	Akebono	Brembo	A+P	Brembo
	Brakes discs	Brembo CCR Carbon Industrie	Carbon Industrie	Brembo	Hitco	Hitco
	Wheels	BBS	Enkey	O.Z.	AVUS	BBS
	Radiators	Secan	Calsonic - IMI	Calsonic	Marston	Nippon - Denso
	Oil tank	middle position inside fuel tank	middle position inside fuel tank	middle position inside fuel tank	middle position inside fuel tank	middle position inside fuel tank
GEARBOX		Longitudinal Carbon	Longitudinal Carbon	Longitudinal Titanium	Longitudinal Titanium	Longitudinal Titanium
	Gear selection	Semiautomatic 7 gears	Semiautomatic 7 gears	Semiautomatic 7 gears	Semiautomatic 6 gears	Semiautomatic 7 gears
	Clutch	Sachs	A+P	A+P	A+P	A+P
	Pedals	2	2	2	2	2
ENGINE		Ferrari 056	Mercedes FO108V	BMW P86/8	Renault RS27	RVX 08
	Total capacity	2400 cmc	2400 cmc	2398 cmc	2400 cmc	2398 cmc
	N° cylinders and V	8 - V 90°	8 - V 90°	8 - V 90°	8 - V 90°	8 - V 90°
	Electronics	Magneti Marelli	McLaren el. sys.	Bmw	Magneti Marelli	Magneti Marellli
	Fuel	Shell	Mobil	Petronas	Elf	Esso
	Oil	Shell	Mobil	Petronas	Elf	Esso
	Fuel tank capacity	90 kg*	98 kg*	98 kg	98 kg*	98 kg
	Dashboard	Magneti Marelli	McLaren	BMW Sauber	Magneti Marelli	Toyota

Car **TABLE**

14 - 15 TORO ROSSO	9 - 10 RED BULL	7 - 8 WILLIAMS	16 - 17 HONDA	20 - 21 FORCE INDIA	18 - 19 SUPER AGURI
STR 03	**R. B. 4**	**FW 30**	**RA 108**	**VJM 01**	**SA 08**
Giorgio Ascanelli Alex Hitzinger-Laurent Mekies	Adrian Newey Geoff Willis•	Patrick Head - Sam Michael Edi Wood	Ross Brawn Jorg Zander	Mike Gascoyne - James Key	MarK Preston Peter McCool
Claudio Balestri (14) Riccardo Adami (15)	Giullaume Rocquelin (9) Ciaron Pilbeam (10)	T. Ross (7) Xevi Pujolar (8)	Andrew Shovlin (16) Jock Clear (17)	Jody Eggington (20) Brand Joyce (21)	Richard Connell (18) Richard Lane (19)
Paolo Piancastelli	Kenny Handkammer	Carl Gaden	Matt Deane	Andy Deeming	Phill Spencer
3140 mm	3140 mm	3159 mm	3192 mm	3097 mm*	3190 mm
1440 mm	1440 mm*	1480 mm	1460 mm	1480 mm	1445 mm
1410 mm	1410 mm*	1420 mm	1420 mm	1410 mm	1405 mm
2+1 dampers and torsion bars	2+1 dampers and torsion bars	2+1 dampers and torsion bars	2+1 dampers and torsion bars	2+1 dampers and torsion bars	2+1 dampers and torsion bars
2+1 dampers and springs	2+1 dampers and springs	2+1 dampers and torsion bars	2+1 dampers and torsion bars	2+1 dampers and springs	2+1 dampers and torsion bars
Koni	Multimatic	Williams	Koni	Sachs	Ohlins
Brembo	Brembo	A+P	Alcon (Brembo)•	A+P	Alcon
Hitco	Hitco	Carbon Industrie	Hitco	Hitco - Brembo	Hitco
O.Z.	O.Z.	O.Z.	BBS	BBS	BBS
Marston	Marston	IMI Marston	IMI Marston - Showa	Secan	Secan - Marston
middle position inside fuel tank	middle position inside fuel tank	middle position inside fuel tank	middle position inside fuel tank	middle position inside fuel tank	middle position inside fuel tank
Longitudinal Alluminium	Longitudinal Alluminium	Longitudinal Titanium	Longitudinal Carbon	Longitudinal Magnesium	Longitudinal carbon/aluminium
Semiautomatic 7 gears	Semiautomatic 7 gears	Semiautomatic 7 gears	Semiautomatic 7 gears	Semiautomatic 6 gears	Semiautomatic 7 gears
A+P	A+P	A+P	Sachs	A+P	Sachs
2	2	2	2	2	2
Ferrari 056	Renault RS27	Toyota RVX 08	Honda RA808E	Ferrari 056	Honda RA808E
2400 cmc	2398 cmc	2398 cmc	2400 cmc*	2398 cmc	2400 cmc
8 - V 90°	8 - V 90°	8 - V 90°	8 - V 90°	8 - V 90°	8 - V 90°
Magneti Marelli	Magneti Marelli	Magneti Marellli	Honda Athena 206	Magneti Marelli	Honda Eneos
Shell	Elf	Esso	Elf	Shell	Eneos
Shell	Elf	Esso	Nisseki	Shell	Eneos
90 kg*	95 kg	98 kg	100 kg	100 kg	100 kg*
Magneti Marelli	Red Bull	Toyota	Bar Honda	P.I.	SA F1

1) From Belgian GP.
2) Retired after Spanish GP.

* Extimated value

The uncontrolled explosion of costs in F1 pushed the International Federation into intervening: in recent years, the organisation has tried to reduce engines to a sort of accessory of the car, curbing research and imposing shared choices to the point that, during 2008, Max Mosley even threatened the future adoption of a single engine. Naturally the constructors were opposed to that, but the regulation restrictions have increased, sterilising important lines of development and leaving, therefore, a situation in which the search for performance only took place through new oils and petrols, aimed at reducing friction.
Pursuing the idea of reducing costs in the engine area, FIA first froze the development of the V8 in 2007, obliging constructors to use a single power unit for two Grand Prix weekends (Friday practice excluded) and then, in 2008, it imposed a certain electronic management system on all the teams. The ruling sparked off enormous controversy, because the winner of the tender for the unit was the McLaren Electronic System, that of the Woking company of which Ron Dennis, who won the drivers' title with Lewis Hamilton, is the boss. The adoption of the MES in the early months of 2008 complicated the lives of the engine specialists: every V8 had been conceived and developed with its own unit, which responded to the specific needs of the respective engines. But many functions on the MES were simplified; traction control and the electronic engine brake had been banned. Hardware had not been designed for the obligatory electronic management system, so they re-used that of McLaren with a set of instructions of limited use. If the technicians of Woking had been able to fully dedicate themselves to the development of the MP4-24, those of the other teams had to hold entire test sessions to make the MES system 'dialogue' with that which managed each V8 from the moment they used very different languages. A complex task, because the unit and its strategy must be calibrated into a system that includes the car, engine and driver. According to unconfirmed reports, the teams were not able to overcome all the problems until mid-season. It should be said that the MES was committed to follow the guidelines of the constructors, so much so that about 50

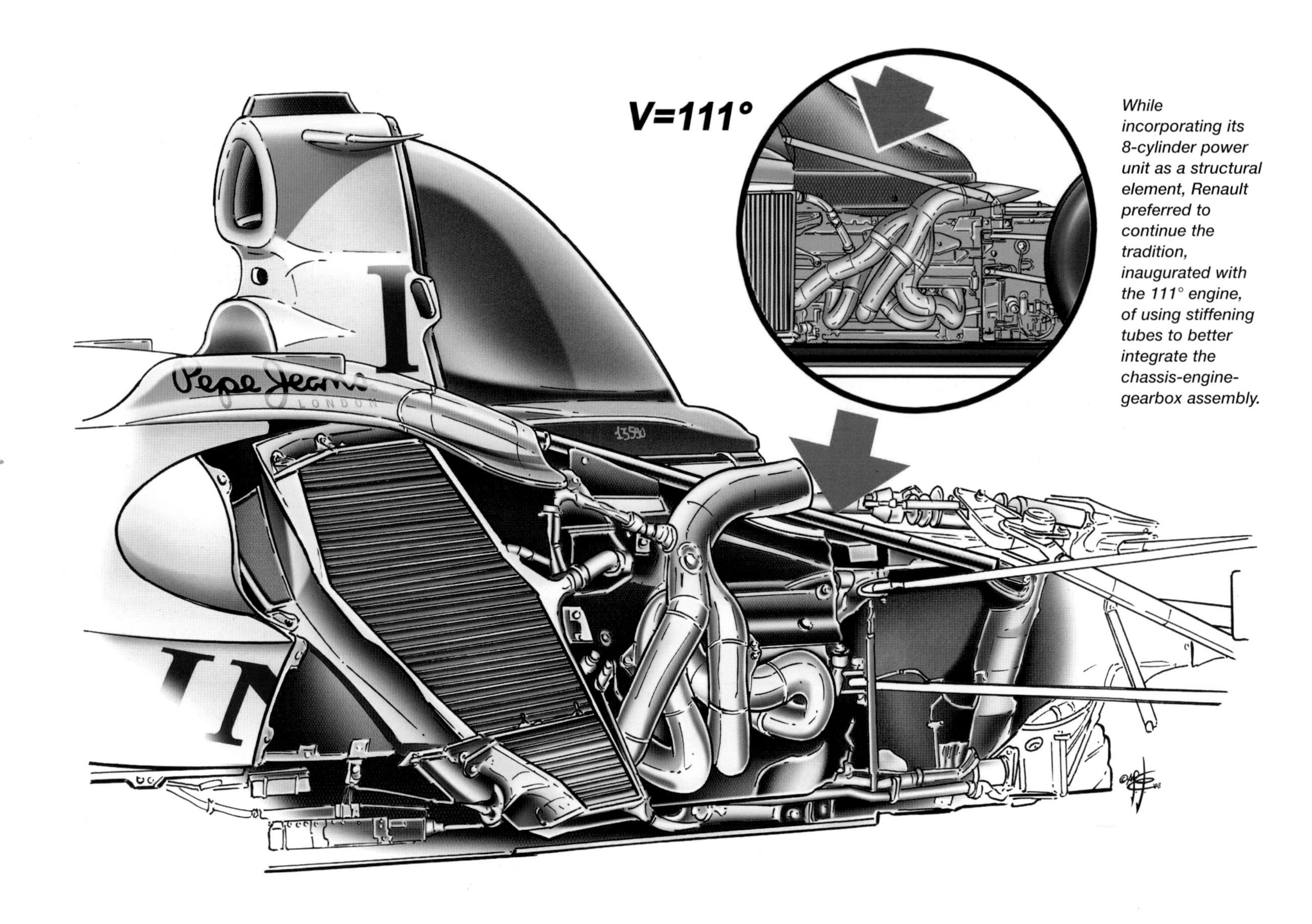

While incorporating its 8-cylinder power unit as a structural element, Renault preferred to continue the tradition, inaugurated with the 111° engine, of using stiffening tubes to better integrate the chassis-engine-gearbox assembly.

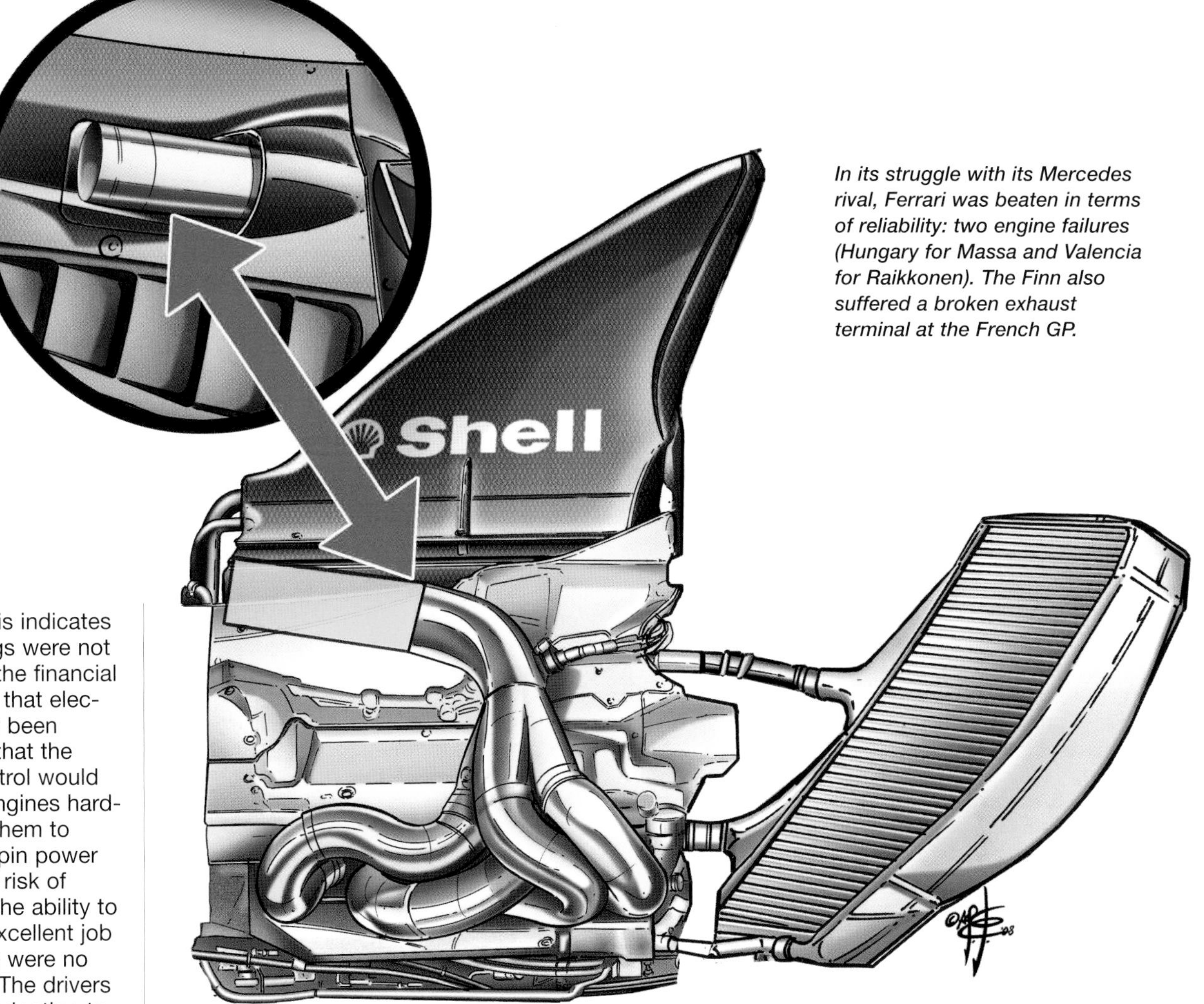

In its struggle with its Mercedes rival, Ferrari was beaten in terms of reliability: two engine failures (Hungary for Massa and Valencia for Raikkonen). The Finn also suffered a broken exhaust terminal at the French GP.

releases were introduced. All of this indicates that the much vaunted cost savings were not made – except by McLaren – but the financial benefit will be seen in 2009, given that electronic development has practically been reduced to nothing. Some feared that the abolition of electronic traction control would lead the drivers to pushing their engines harder under acceleration, subjecting them to harmful effects during the wheel spin power stage on exiting a corner, with the risk of over-revving. It was either due to the ability to adapt on the drivers' part or the excellent job done by the technicians, but there were no breakdowns due to that problem. The drivers said they had greater difficulty in adapting to the lack of the active engine brake. In 2007, the use of engine braking was optimised electronically to have the correct split between the front and rear brakes, avoiding wheel blocking damage in the second part of the disengagement, varying the opening of the butterfly, anticipating the ignition and the variation of phase. But in 2008, the MES no longer allowed the link-up of the engine brake map to the speed of the car for which reason cars like the Toyota, with its unstable rear end under braking, was even more difficult for the drivers to handle (especially Trulli). Due to this important variation, FIA allowed the constructors to adjust the V8s to guarantee indispensable reliability, it being prohibited to go in search of more performance.
Ferrari, who won the constructors' title, had to surrender the drivers' championship precisely for reliability problems of the V8 – the 056 – especially early on in the year, conditioning the fantastic season of Felipe Massa, who fought for the title with Lewis Hamilton until the last corner. The double KO of the F2008s in Australia was the alarm bell, because such a thing had not happened since the 1994 GP of Belgium. The most striking case took place in Hungary, when Massa retired on the last lap while he was firmly in the lead: what broke was a connecting rod that had missed out on some treatment by Pankl, the trusted Austrian supplier to the Prancing Horse. To improve the aerodynamic efficiency of the Rossa, without shadow of doubt the best car of the championship, the engine specialists had put their money on the reduction of the heat mass, making more air pass at greater speed along the sides. But because the 056 was a 'hot' engine – the temperature of the water was about 125 °C – it overheated during GPs held in extremely hot weather: on some occasions, the telemetry data advised a reduction in revolutions – the limit was 19,000 rpm – to stop the unit from breaking down. It should be said that the Mercedes-Benz V8 was no stranger to breakdowns, but McLaren's 'good fortune' was that the problems only hit Heikki Kovalainen (Japan and China) sparing championship leader Hamilton. It is also worth recalling Raikkonen's GP of France: the Finn was dominating Magny-Cours when one of his exhausts that protruded from the engine cover broke. The Ferrarista complained of a loss of power (vibration broke the welding of a terminal, which had been further lightened) but the men at the pit wall directed by Luca

Baldisserri were quick to give Kimi useful instructions so that he could carry out the indispensable regulation on the steering wheel-computer and could complete the race in second place despite around a 50 hp drop in power!
Even if the first engine break of 2008 did not cost 10 places on the starting grid, as happened in 2007, the technicians went about developing strategies that could protect the engine in view of the second race. It is no mystery that cars register a loss of power during the course of a GP and the reduction in performance is considerable if very high operating temperatures are used. One can reasonably talk of power loss values of between 0.5 and 1.5%: about 7 hp per GP which, in the second race, could reach 15 hp under stress conditions. According to the statistics, the most reliable engine was the BMW, ahead of the Toyota and Mercedes, but it is reasonable to ask which was the most powerful. Obviously official data does not exist, but estimates from phonometric checks of the various teams do. And here there emerges some very interesting information that should be interpreted, because there are V8s that have been developed to produce a good filling of the power curve at the cost of maximum power, and others whose technicians went in search of horse power. Mercedes and Toyota were accredited with 730 hp tops, while the 056 Ferrari was just below that and BMW was accredited with 730 hp, Honda with 720 hp and Renault were relegated to around 710 hp. But this is not the data that testifies to the good quality of a unit. Certainly, it is more interesting for a specialist to verify how his engine conducts itself in the medium range (16,000-19,000 rpm), which is of greater use. Excellence distinguishes Ferrari, Mercedes and Toyota with 730/740 hp, with BMW very close to 720 hp, while Honda and Renault were more distant. The French constructor, which had more or less budgeted half the amount of money of the others in the engine area, began the season with a V8 of 670 hp in this range of use with serious reliability problems. FIA authorised work on the engines of Viry-Chatillon and the power increase, which still did not take the French motor to the top, enabled Fernando Alonso to win two GPs. Despite the 'freeze' restrictions, the engine specialists were able to gain 10 or so horse power during 2008: this was as a result of the reduction of friction by working on lubricant oils – their density is now of extremely low viscosity – and petrol developed especially for each circuit in relation to the characteristics of the track and the climatic conditions. And that is the proof that progress is not stopped by regulations...

Franco Nugnes

Race retirements due to engine breakages

Unit	Driver	Kms	No.race	Race
Ferrari 056 (3)	S. Bourdais	434	1	Australia
Ferrari 056 (2)	F. Massa	318	1	Australia
Ferrari 056 (1)	K. Raikkonen	387	1	Australia
Ferrari 056 (4)	S. Vettel	578	2	Malaysia
Renault RS27 (8)	F. Alonso	787	2	Spain
Toyota RVX-08 (7)	N. Rosberg	868	2	Spain
Ferrari 056 (33)	F. Massa	900	2	Hungary
Ferrari 056 (37)	K. Raikkonen	934	2	Europe
Mercedes FO108V (17)	H. Kovalainen	689	2	Japan
Mercedes FO108V (19)	H. Kovalainen	441	1	China
Ferrari 056 (35)	S. Vettel	784	2	Hungary

Average kms per engine model

Engine	Units used	Kms covered	Av. kms per unit	Team users
BMW P86/8	19	16.953	892	BMW Sauber
Toyota RVX-08	39	32.194	825	Toyota, Williams
Mercedes FO108V	20	15.635	782	McLaren
Renault RS27	40	28.899	722	Renault, Red Bull
Honda RA808E	24	17.241	718	Honda, Super Aguri
Ferrari 056	62	42.919	692	Ferrari, Toro Rosso, Force India

In the absence of substantial modifications to the regulations, from the engine point of view 2008 was a logical continuation, if not a photocopy, of the previous season. The average use from the kilometrage point of view remained more or less unchanged, as was the scale of values, which were almost identical to those of 2007.

For the second year in succession, BMW is confirmed as the engine that covered most road, followed by Toyota and Mercedes-Benz, although the retirements of Kovalainen in Japan and China took their toll.
At the bottom of the classification once again was the Ferrari 056 which, as in 2007, was the power unit with the least average distance covered; a negative award to which it also came close in 2006, when it was second last to the Cosworth CA2006, a company that retired from the sport at the end of that season.

Worrying data for the Rosse, who literally saw a significant victory go up in smoke in Hungary when an engine blew, forcing Massa to retire just a few kilometres from the finish line. The fragility of the 056 was confirmed by the fact that out of 11 retirements from races due to engine breakdowns during the season, no fewer than seven were Maranello's units.

The only data that varied significantly compared to 2007 was that of penalisation for the substitution of engines.
The regulation variation that allowed a change without penalty meant that throughout the entire season only Webber in China was hit by what was once the feared regression of 10 places on the grid, a problem that also cost Kimi Raikkonen a large slice of the title in 2005. Even without the regulation alteration, engine changes would still have been very limited with only eight throughout the season, data in further decrease compared to the surprising 11 of 2007.

On the record side there was also continuity: Nick Heidfeld was at the top for average kilometrage and it was his engine that covered most distance: it was the unit used in Belgium and Italy at 1,094 kilometres.

Average kms covered per team

Team	Engine	Unit used	Total kms	Average kms/engine
BMW Sauber	BMW P86/8	19	16953	892
Williams	Toyota RVX-08	18	15857	881
McLaren	Mercedes FO108V	20	15635	782
Toyota	Toyota RVX-08	21	16337	778
Ferrari	Ferrari 056	22	16100	732
Honda	Honda RA808E	20	14548	727
Renault	Renault RS27	20	14493	725
Red Bull	Renault RS27	20	14406	720
Toro Rosso	Ferrari 056	20	14404	720
Super Aguri	Honda RA808E	4	2693	673
Force India	Ferrari 056	20	12415	621

Engines with the most km

Driver	Unit	Kms covered	Used in
N. Heidfeld	BMW P86/8 (14)	1094	Belgium + Italy
T. Glock	Toyota RVX-08 (29)	1086	Belgium + Italy
S. Bourdais	Ferrari 056 (40)	1067	Europe + Belgium
S. Vettel	Ferrari 056 (41)	1030	Europe + Belgium
J. Trulli	Toyota RVX-08 (5)	1029	Malaysia + Bahrain
N. Rosberg	Toyota RVX-08 (27)	1020	Belgium + Italy
R. Kubica	BMW P86/8 (13)	1019	Europe + Belgium
S. Vettel	Ferrari 056 (48)	1013	Italy + Singapore
S. Bourdais	Ferrari 056 (54)	1009	Japan + China
T. Glock	Toyota RVX-08 (25)	1007	Hungary + Europe

Engines with the least km

Driver	Unit	Kms covered	Used in
S. Vettel	Ferrari 056 (11)	124	Bahrain
D. Coulthard	Renault RS27 (15)	140	Monaco
A. Sutil	Ferrari 056 (5)	143	Australia
G. Fisichella	Ferrari 056 (19)	149	Turkye
N. Piquet	Renault RS27 (41)	159	Brazil
F. Massa	Ferrari 056 (46)	202	Italy
A. Sutil	Ferrari 056 (7)	241	Australia + Malaysia
J. Trulli	Toyota RVX-08 (3)	249	Australia
N. Piquet	Renault RS27 (2)	259	Australia
D. Coulthard	Renault RS27 (3)	302	Australia

(the progressive number of units produced by the engine manufacturer in brackets)

Engines used during the season

Driver	Unit	Used
Kazuki Nakajima	Toyota RVX-08	9
Lewis Hamilton	Mercedes FO108V	9
Nick Heidfeld	BMW P86/8	9
Nico Rosberg	Toyota RVX-08	9
Adrian Sutil	Ferrari 056	10
David Coulthard	Renault RS27	10
Fernando Alonso	Renault RS27	10
Giancarlo Fisichella	Ferrari 056	10
Jarno Trulli	Toyota RVX-08	10
Jenson Button	Honda RA808E	10
Mark Webber	Renault RS27	10
Nelsinho Piquet	Renault RS27	10
Robert Kubica	BMW P86/8	10
Rubens Barrichello	Honda RA808E	10
Sebastian Vettel	Ferrari 056	10
Sebastien Bourdais	Ferrari 056	10
Felipe Massa	Ferrari 056	11
Heikki Kovalainen	Mercedes FO108V	11
Kimi Raikkonen	Ferrari 056	11
Timo Glock	Toyota RVX-08	11
Anthony Davidson	Honda RA808E	2
Takuma Sato	Honda RA808E	2

Other statistical curiosities:

- *the two contenders for the title, Hamilton and Massa, were separated by 140 kilometres of average coverage and two engines used, to the disadvantage of the Brazilian.*
- *the difference between the Ferrari 056 that powered the Maranello cars and those of Force India was notable: an average of 732 km for the Ferrari and 621 – last place – for the cars of Fisichella and Sutil.*
- *a double retirement of the two Ferraris took place in Australia due to engine trouble, the first since the 1994 GP of Belgium.*
- *the last time the race leader's engine has broken in recent seasons was Nick Heidfeld's during the 2006 GP of Malaysia.*

Average kms per engine model

Driver	Engine	Used	Total kms	Average kms/engine
Nick Heidfeld	BMW P86/8	9	8.626	958
Nico Rosberg	Toyota RVX-08	9	8.231	915
Lewis Hamilton	Mercedes FO108V	9	7.853	873
Kazuki Nakajima	Toyota RVX-08	9	7.626	847
Robert Kubica	BMW P86/8	10	8.327	833
Jarno Trulli	Toyota RVX-08	10	8.123	812
Fernando Alonso	Renault RS27	10	7.947	795
Mark Webber	Renault RS27	10	7.744	774
Takuma Sato	Honda RA808E	2	1.548	774
Timo Glock	Toyota RVX-08	11	8.214	747
Rubens Barrichello	Honda RA808E	10	7.397	740
Sebastien Bourdais	Ferrari 056	10	7.364	736
Felipe Massa	Ferrari 056	11	8.093	736
Kimi Raikkonen	Ferrari 056	11	8.007	728
Jenson Button	Honda RA808E	10	7.151	715
Heikki Kovalainen	Mercedes FO108V	11	7.782	707
Sebastian Vettel	Ferrari 056	10	7.040	704
David Coulthard	Renault RS27	10	6.662	666
Nelsinho Piquet	Renault RS27	10	6.546	655
Giancarlo Fisichella	Ferrari 056	10	6.495	650
Adrian Sutil	Ferrari 056	10	5.920	592
Anthony Davidson	Honda RA808E	2	1.145	573

The most important new technological development of the 2008 season was the banning of traction control and a number of other electronic driver aids. So anti-spin, introduced in 1990, was unable to come of age due to its abolition. It was Ferrari that brought in the device with the 641, as can be seen in the illustration, bringing its considerable influence to bear on driving styles. A new feature that arrived just a year later from the great revolution was the semiautomatic gear management on the steering wheel, which was also introduced once again by Ferrari in 1989 on John Barnard's 639.
With the 2008 season, the Federation took a step back in relation to increasing the value of the drivers' driving style. It decreed a complete clamp down on sophistication guaranteed by electronics, imposing a single electronic management system on the car. The unit was a little backward, but it was the same for everyone. It did have a series of limitations compared to the most sophisticated electronic management systems. Supplied by a company belonging to McLaren called MES, it created some discontent and problems among the other teams at the start of the season. Ferrari had complained about the lesser sophistication of the unit compared to its Magneti Marelli component, especially at the opening race of the season in Melbourne. It ended traction control, assisted starts, reading the position of a car in relation to the various corners of a circuit to optimise handling with specific mapping; all electronic aids that had minimised the capabilities of the drivers too much.
Without electronic control of the generation of power, drivers had to find throttle pedal sensitivity once more, returning to dosing the pedal with caution and skill after years of using more or less brutal efforts with the safety net of traction control, always ready to avoid trouble, skidding most of all. The effect was seen right from the first test: side-slipping had returned and the errors increased, all to the advantage of spectacle, which was guaranteed for the whole season with a heart-stopping finale at the last race in Brazil. The drivers, but also the cars, had to adapt themselves to new requirements and designing the latter included the need of ensuring greater mechanical control by the pilots. In preparing for the 2008 season, the teams worked a great deal on engine power output, with drivers obliged to choose the travel of the accelerator pedal. Unquestioned penalisation came from the abolition of the engine brake, which permitted a slight correction of power in breaking away on the limit to avoid blockage of the rear wheels. This was an operation that was even more difficult to achieve with only the sensitivity of the right foot and it had a determinate influence on the brake balance. During the 2008 season, the role of the driver was more decisive, too. We finally saw corrections of errors, sometimes imperceptible, but which played their part in performance in the end. This factor often complicated the life of those who had to study race strategies and fuel quantities taken on board for the last qualifying stint. In fact, very often the reason for a difference in performance between Q2 and Q3 was not just due to the different amounts of petrol on board, but also the sum of small errors that influenced the performance of the driver.

McLAREN STEERING WHEEL

Right from the first 2008 race it appeared evident that, with the abolition of electronic management of traction control and the engine brake certain functions, which were previously managed automatically by the system, were now required by manual management by the driver using steering wheels that were much more sophisticated. The most evident new development was the one brought in by McLaren, with a wheel that looked as if it was completely identical to the one of the previous season from the front, but which showed a brand new feature – the presence of a further couple of paddles (1) as well as the usual ones for the change of gears (2) and clutch (3), the number becoming no fewer than six instead of the usual 4. That way, Ron Dennis's team had partially got around the lesser versatility of the MES electronic management system by reproducing the anti-skid effect manually. The new paddles managed different maps linked to engine torque and they were used jointly in the various changes of gear (arrow to the right), uniting in one fell swoop the control of the two functions. That is why these added levers were fitted close to those of the gears, so close that it made individualisation difficult.

2008 REGULATIONS

GEARBOX FOR 4 RACES

The obligation was confirmed for the use of the same engine for two races with the only new feature a bonus of no penalisation of 10 positions on the grid for the first breakdown of the season. But another new rule was introduced. It was the requirement to use the same gearbox for four consecutive races. This was introduced for the purpose of reducing costs and carried a penalty of 5 grid places in the case of infraction. It should immediately be clarified that this rule only concerned the actual gearbox and the principal shafts. Teams were, obviously, permitted to change gear ratios from one track to another, as they were to replace the same gears in the case of manifest breakage on the authorisation of the Federation's technical delegate. In addition, a minimum limit of 12 mm was fixed for the thickness of the ratios in a further effort to reduce costs.

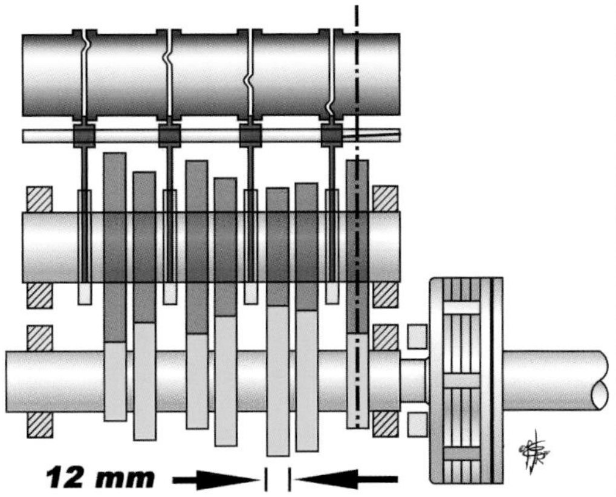

12 mm

ABOLITION OF THE SPARE CAR

There were no spare cars in the pits of the various teams during the 2008 Grands Prix, leaving a lot of extra space in which the mechanics could work. The teams could bring a third monocoque, but it had to be unassembled (the one that was the fourth chassis available until 2007 in case of accidents that required the original's replacement). In practice, if a driver ruined the monocoque of his car he could no longer participate in the practice of that day, partly because the operation of the complete assembly of the third monocoque was authorised only for the following day. A naked chassis in the past could be assembled in not much more than 6 hours, so in time to get the driver who had damaged his car in the first practice of the morning back on the track for the last stint of the afternoon. However, the Federation purposely impeded such a race against time, in part to place a limit on the number of mechanics in the various teams. A rule that could become serious in the end by excluding a driver involved in an accident on the Saturday morning from qualification in the afternoon, as happened to Raikkonen at Monza. The absence of the spare car penalised the task of observers: in recent seasons, the latest modifications were sometimes not applied to the third car assembled by the teams, the updates being reserved for the two cars for the race. That facilitated the job of those who wanted to discover the differences.

SPARE CHASSIS

Until the 2007 season, the spare fourth monocoque was transported to the circuits pre-assembled, with all its cabling and radiator packs: in some cases, they were even transported with the front suspension already fitted. But the 2008 regulations dictated that assembly had to begin from a completely naked monocoque.

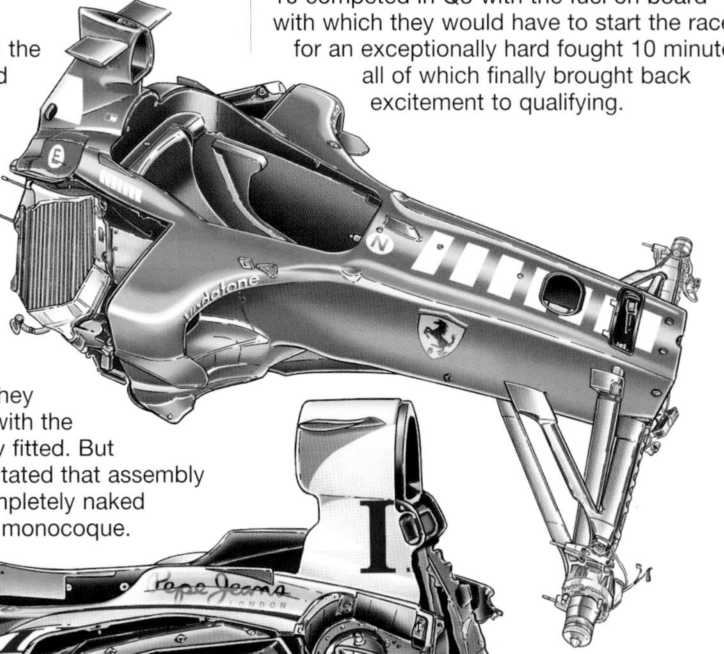

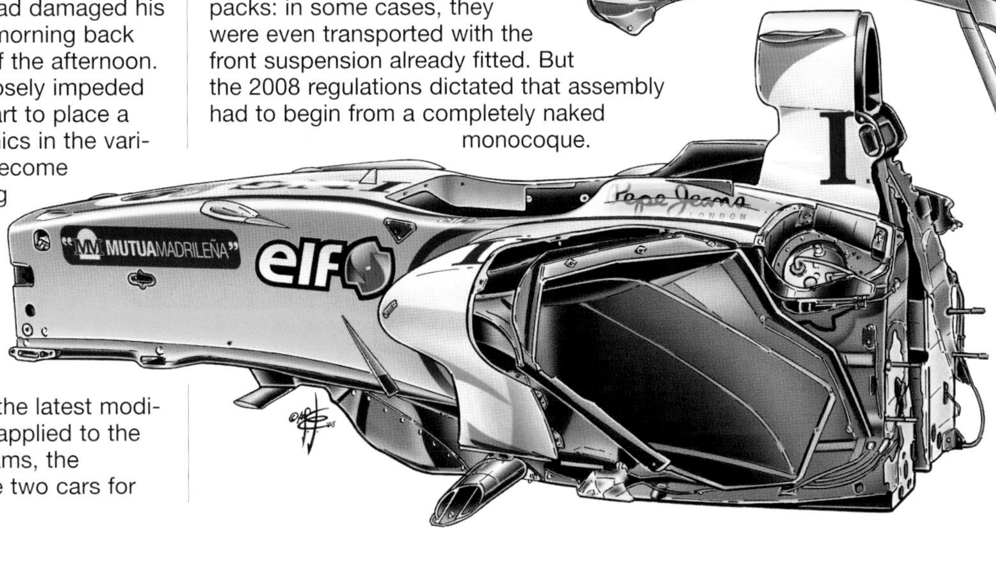

TYRES

The obligation to use both types of tyres supplied by Bridgestone for the race was maintained, as was marking the soft version in white, which was also extended to the 'full wet' among those versions to facilitate recognition. Fourteen sets of dry weather tyres were available to every competitor (2 of each individual type, usable in free practice), 4 wet and three full wet.

In practice, the abolition of electronic aids made the task of preserving the tyres more delicate throughout the duration of a race, rewarding the more sensitive drivers and the cars that were less severe on their tyres, with the possibility of creating surprises in the speed values shown in qualifying.

QUALIFYING

The means of qualifying was modified, the most important change concerning the abolition of topping up fuel in the third phase (Q3), cut to just 10 minutes. The duration of the first period (Q1) was also changed to 20 minutes after the elimination of the six slowest cars. The remaining 16 cars had 15 minutes for the second phase (Q2). After that, the best 10 competed in Q3 with the fuel on board with which they would have to start the race for an exceptionally hard fought 10 minutes, all of which finally brought back excitement to qualifying.

COMPARISON: FERRARI STEERING WHEELS

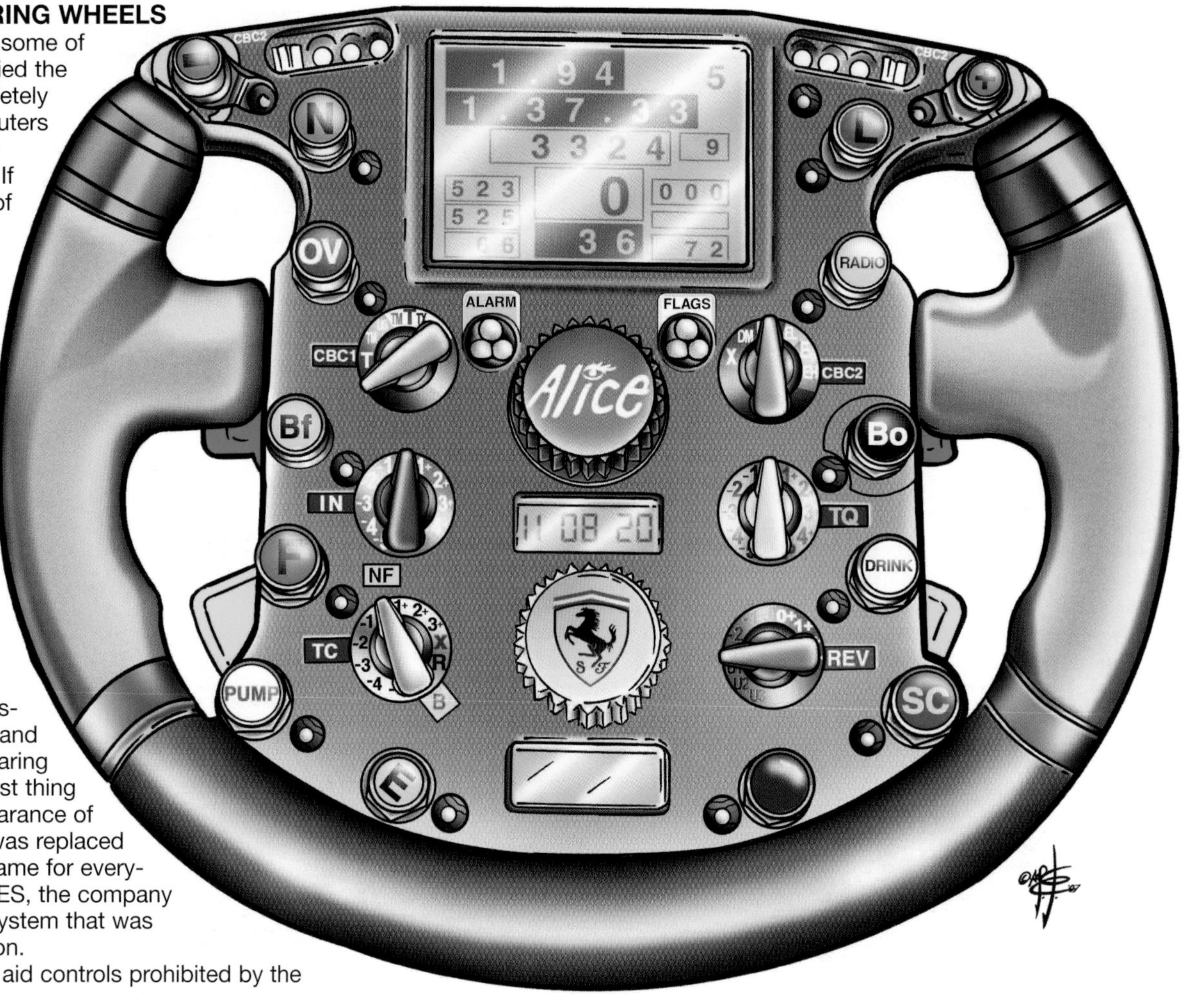

Whoever thought that the abolition of some of the electronic aids would have simplified the steering wheels of F1 cars was completely mistaken. In fact, their onboard computers became much more sophisticated, as described in the chapter on cockpits. If on the one hand the steering wheels of the 2008 cars were partially simplified a series of notable functions and, therefore, paddles having disappeared, on the other hand new ones were added to partly do the job of those eliminated. They had different functions but, as always, were there to help the driver. One example that can represent them all was the new 'tyre' paddle, which indicates the different choices of power generation and the management of the gearbox on the basis of the various situations of the car's tyres. We see from the example of the Ferrari steering wheel, which was the most sophisticated and complete of management, no fewer than 24 paddles and buttons, a liquid crystal central display that seemed like a plasma video and another two smaller auxiliaries. Comparing the 2007 version with the 2008, the first thing that hits one in the eye is the disappearance of the big multi-function display, which was replaced by a more spartan unit that was the same for everyone. The new one was supplied by MES, the company that produced the electronic control system that was imposed on all teams by the Federation.
The red cross indicates the electronic aid controls prohibited by the Federation for 2008:

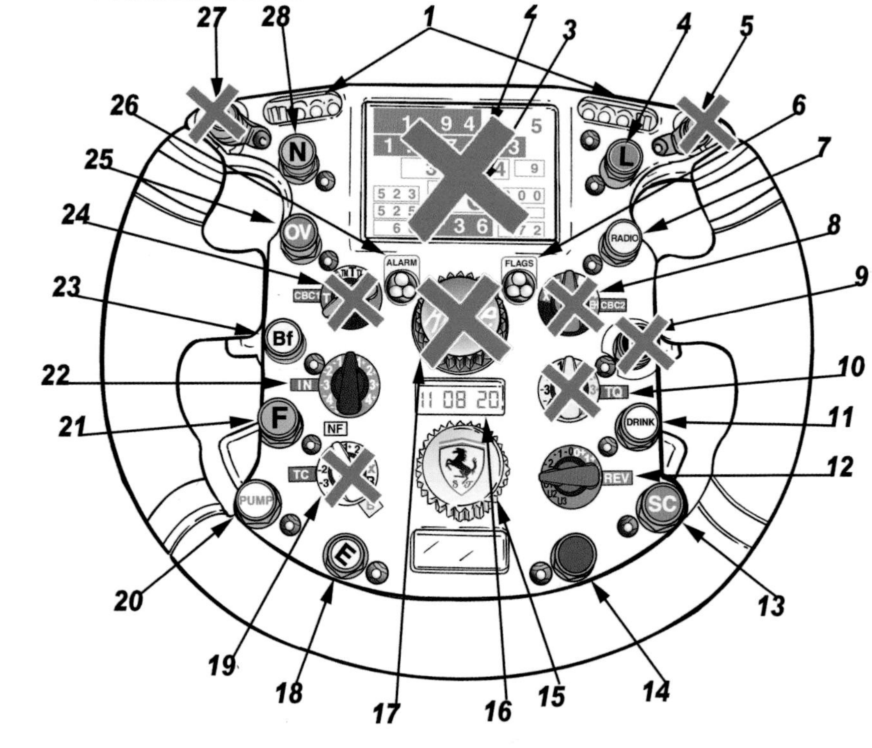

1) Lights in sequence to indicate the correct revolutions at which to change gear.
2) Multi-function central display, showing the lap time.
3) Display that indicates the gear engaged.
4) Pit lane speed limiter.
5-27) enables the driver to move up or down the map of corners.
6) Lights that repeat the trackside signal flags.
7) Radio.
8) Calibration of the differentials, corner by corner.
9) Burn out: its task is to carry out the 'ungum' instruction to bring the tyres up to operating temperature in the warm-up lap.
10) Keeps a check on engine torque.
11) Drink button.
12) Controls the engine's revolutions.
13) Strategy in operation for the safety car.
14) Spare button.
15-17) Paddles that change the screening on the display.
16) Clock.
18) Stops engine.
19) Traction control.
20) Auxiliary petrol pump.
21) Manual control of the fuel door.
22) Traction control on entering the corner.
23) Bit finder: searches for the optimum point for the disengagement of the clutch.
24) Various settings of the differential corner by corner.
25) Over revs: allows the engine to exceed normal maximum revolutions to overtake.
26) Alarm on the track.
27) See 5.
28) Neutral.

2008 STEERING WHEEL CONTROLS

Despite the abolition of many functions described in the captions of the black and white illustration, the 2008 steering wheels remained highly complicated; in some cases they were even given new management paddles. They included 'tyre' in the low centre, which should indicate the different choices of power generation and gearbox management on the basis of the various tyre situations in both practice and the race.

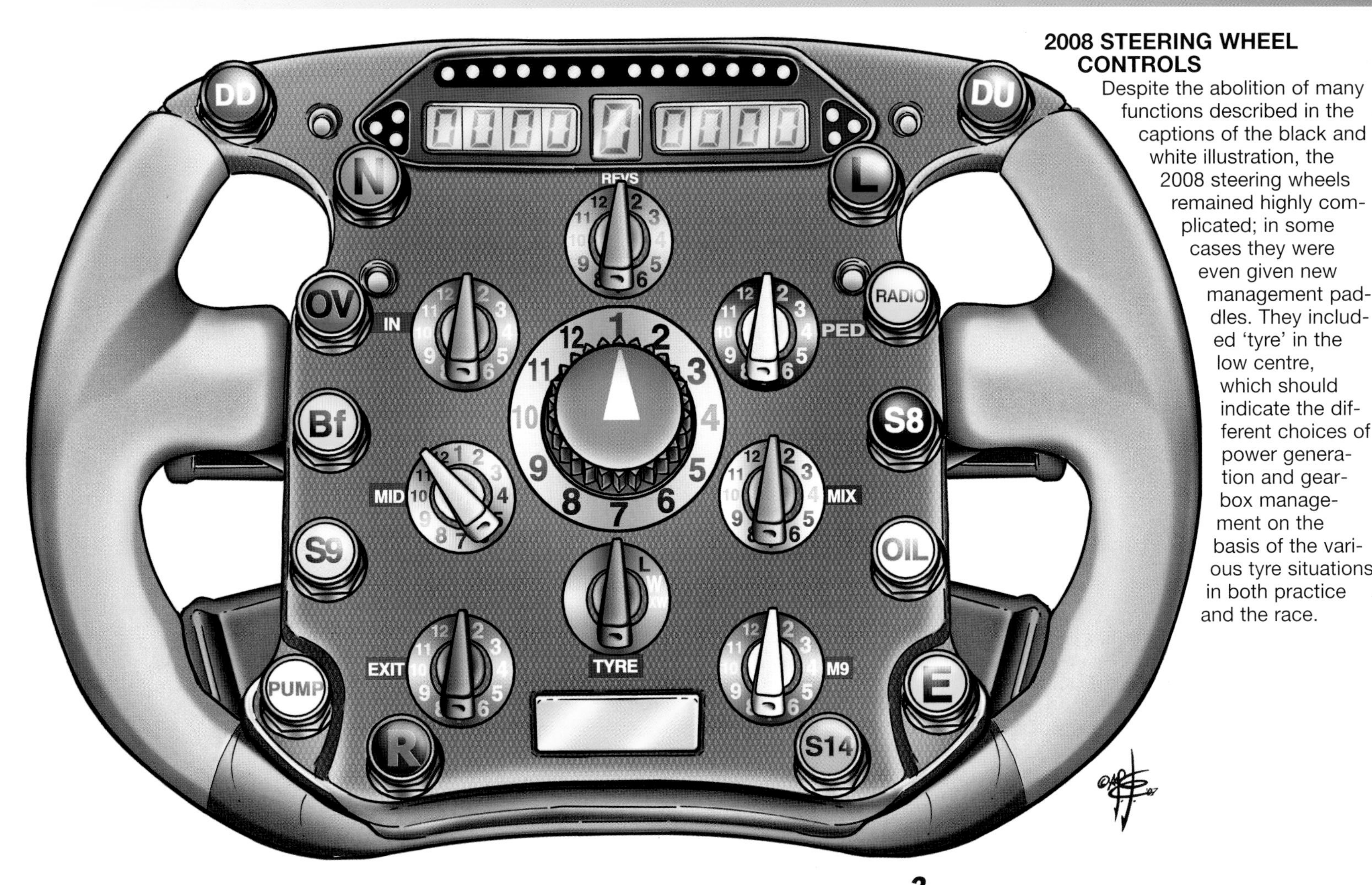

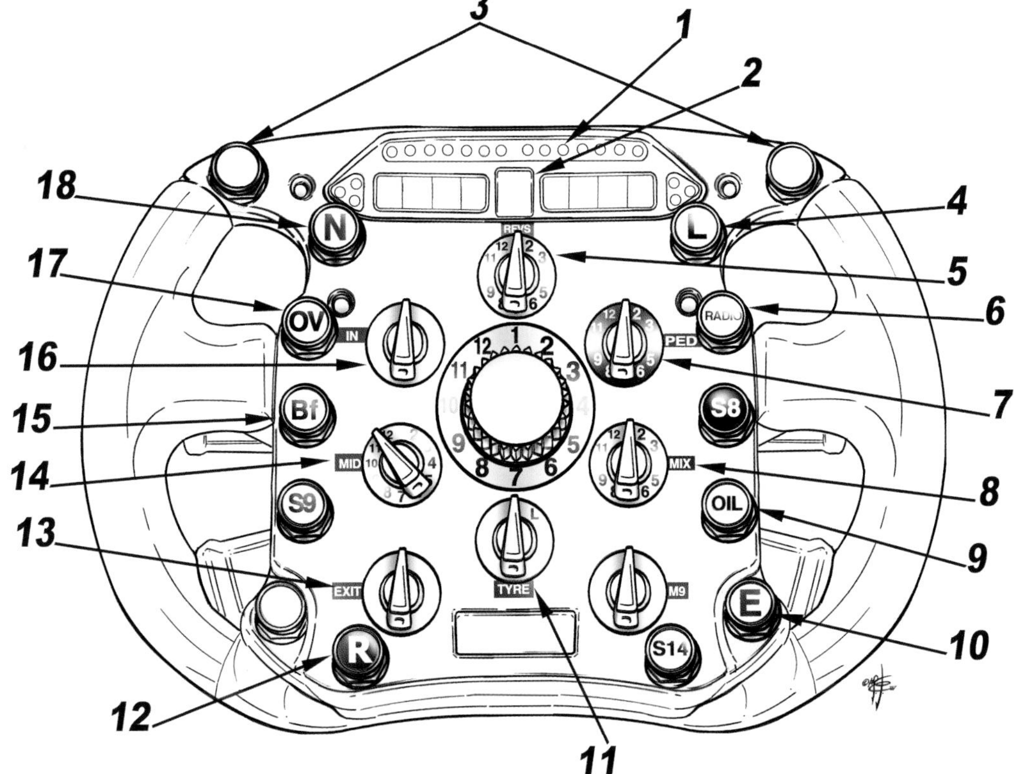

1) Sequential lights to indicate the right revolutions of the engine at which to change gear.
2) Multi-function central display showing the gear engaged in the centre.
3) Button that enables the driver to move from various elements on display.
4) Pit lane speed limiter.
5) Different levels of engine revolutions.
6) Radio.
7) New button that manages accelerator pedal response.
8) Controls carburetion.
9) Operates the auxiliary oil pump.
10) Stops engine.
11) New button that manages power generation in relation to the tyres fitted.
12) Radio confirmation.
13-14-16) Paddles that operate the control of the differential singly when entering, at the centre and exiting a corner.
15) Searches for the clutch contact point.
17) Increases engine revolutions to overtake.
18) Neutral.

As if they were attempting to exploit the rules stability in the field of aerodynamics prior to the advent of the major revolution imposed by the Federation for 2009, the various teams introduced a notable number of new features in the 2008 season, so many in fact that we have decided to expand this chapter, incorporating within it the pages usually dedicated to new features. The novelties of the 2008 season were in fact all concentrated in this sector. The interesting thing is that it was not just the leading teams, Ferrari and McLaren who were setting trends, but other teams too, first and foremost, BMW. The German team was influential above all with two features that debuted on the 2008 car: the vertical link in front of the sidepods between the baffles and the "Viking horns" attached to the nose; the first was copied by all the teams, except for McLaren and Force India, while the second was also adopted by Red Bull and Toro Rosso. Then there was Red Bull's engine cover large fin, which was also adopted by all apart from McLaren, BMW and Williams; curiously, BMW had actually been the first team, back when it was still Sauber (2005), to introduce a dorsal fin. An interesting aspect is that some of these new features are revisitations of innovations introduced in the past and revised to suit the current cars. The sail-like engine cover of the Red Bull, for example, had a forerunner as far back as 1972 with the Ferrari B2 and another in 1995 with the McLaren MP4-10. Even the season's major innovation, the "famous" slot in the nose of the Ferrari had distant origins in the "walrus" design introduced by Williams in 2004. Honda's elephant's ears, later copied by McLaren, had also already been seen during testing for the 2007 season, but had never previously been raced.

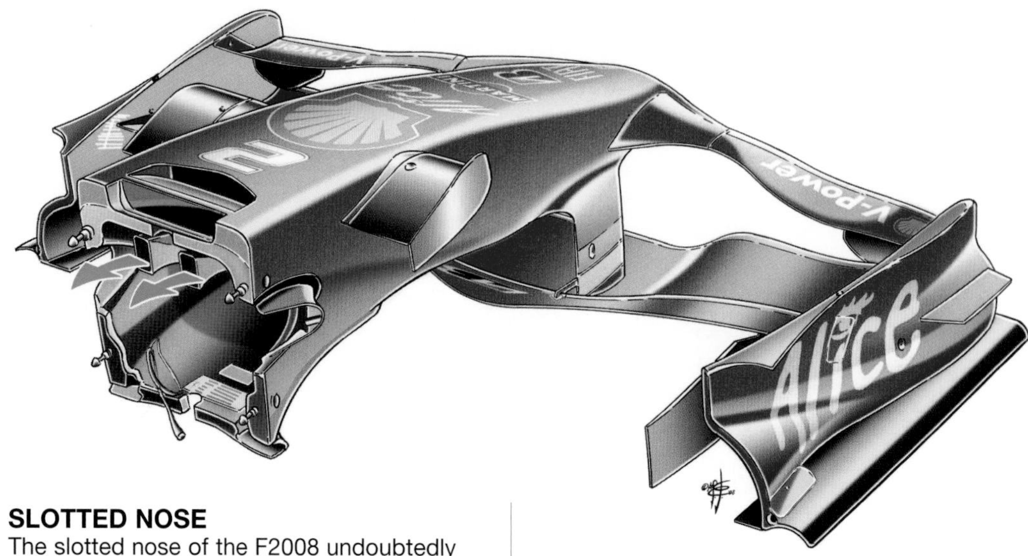

SLOTTED NOSE

The slotted nose of the F2008 undoubtedly represented the most innovative and original technical feature of the season, reprising Williams' unsuccessful and controversial "walrus" design from 2004.
An air intake slot was let into the lower part of the traditional nose, the air flowing through a suitably shaped channel equipped with an internal flap to the upper part of the nose cone itself, joining with the air flows from the front section of the car.
The effects of this feature were two-fold:
1 – increasing the velocity of the air flow under the car, with the air being in part "sucked" upwards where it rejoined the flows around the upper section of the nose; the effects can be seen in the cockpit area and also around the engine air intake. It was possible to adjust the upper opening to vary the flow of aspirated air and therefore the loading on the front of the car.
2 – reducing the quantity of air directed to the lower part of the nose and therefore beneath the bottom of the car. In this way, all the air passing beneath the car can be extracted correctly and can generate the required aerodynamic load. In this way there is less risk of an extractor "stalling" that would reduce its efficiency.

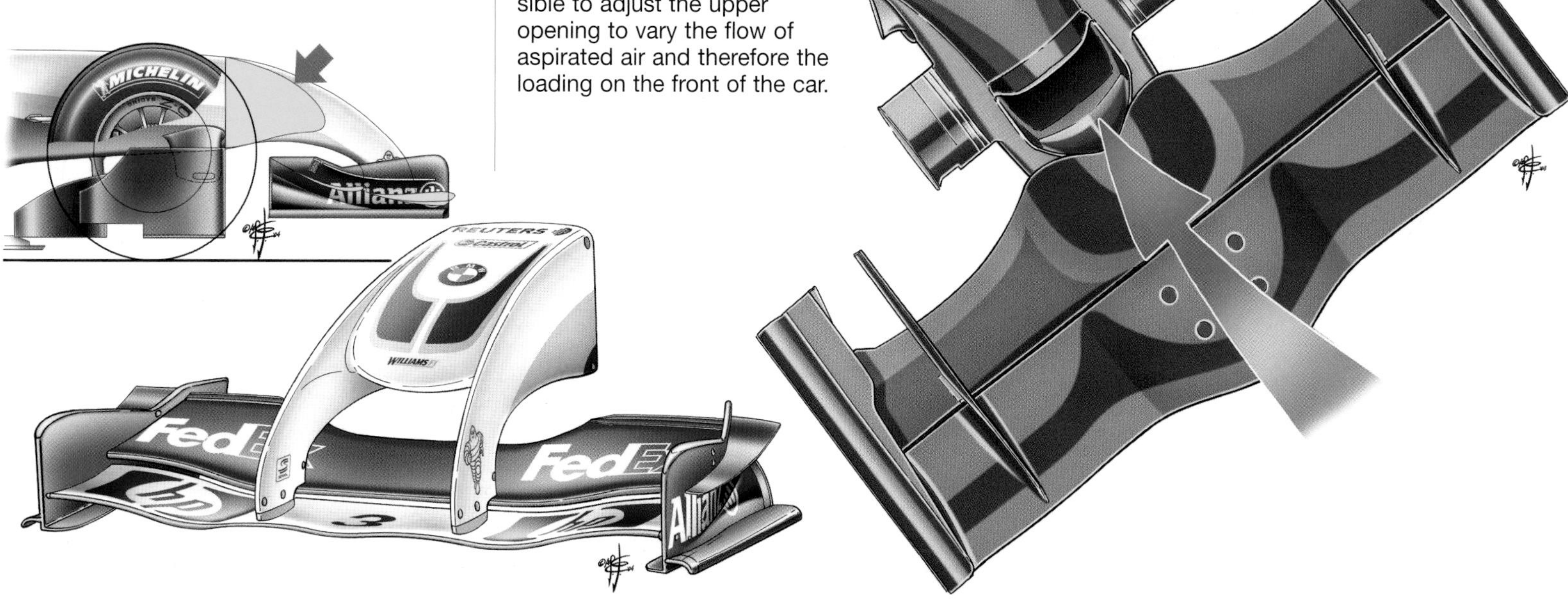

Talking about AERODYNAMICS

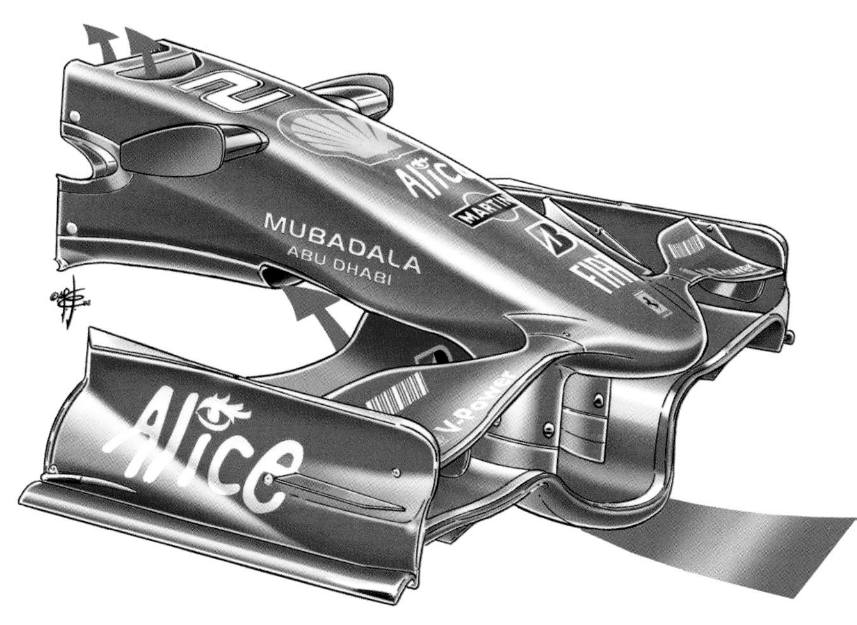

The device was mainly used on tracks with medium/hight aerodynamic load and permitted a reduction in the angle of incidence of the front and rear wings as the downforce generated by the car's underbody had been increased. The difference in maximum speed on tracks like Barcelona was in the order of 5/6 kph.

McLAREN

McLaren's response to Ferrari's slotted nose was to multiply its profiles to four with the addition of a further slot in the central section on ultra-high loading circuits effectively increasing the number of profiles in that area to five. The aim of these slots was to prevent the profiles from stalling when high angles of incidence were employed. The comparison drawings show that the first profile had an extremely reduced chord and that the leading edge was higher than the trailing edge.

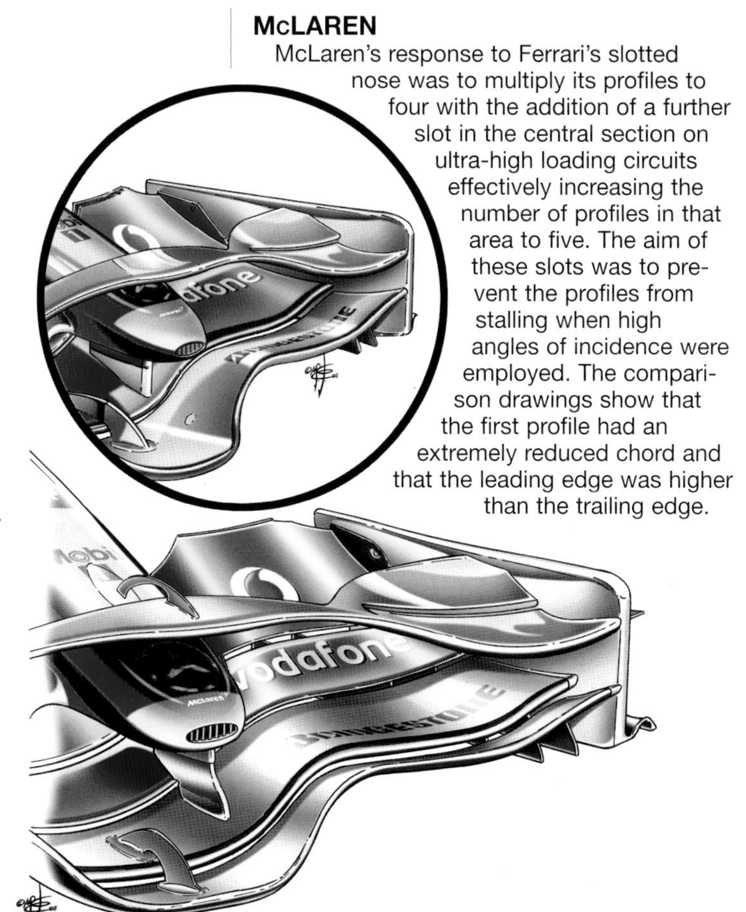

BMW

BMW was responsible for introducing one of the most widely copied features – this element linking the boomerang close to the sidepod and the bargeboards alongside the chassis, ahead of the sidepods themselves. The aim of this feature was to direct the air flows coming from the front end of the car along the sidepods. The air that was neither directed beneath the underbody nor swallowed by the sidepods was first directed parallel to the chassis before being channelled between chassis and bargeboard and finally flowing over the outside of the sidepods; here, this vertical element deviates the flow so that it returns to converge towards the end section of the sidepods in the area of the upper part of the diffuser. Moreover, the position of this element downstream of the front wheels also permits a slight reduction in the turbulence they generate. The "cleaner" this air, the more efficient the rear wing profiles will be.

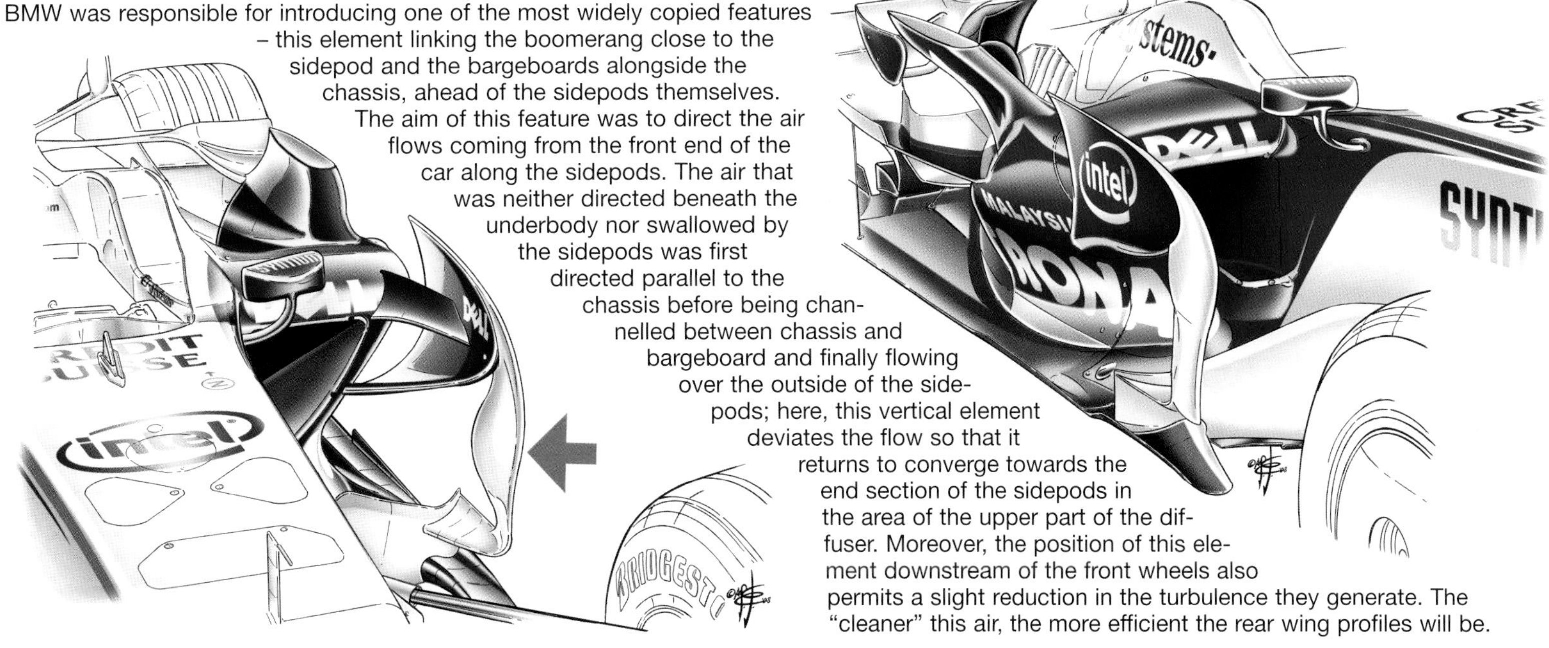

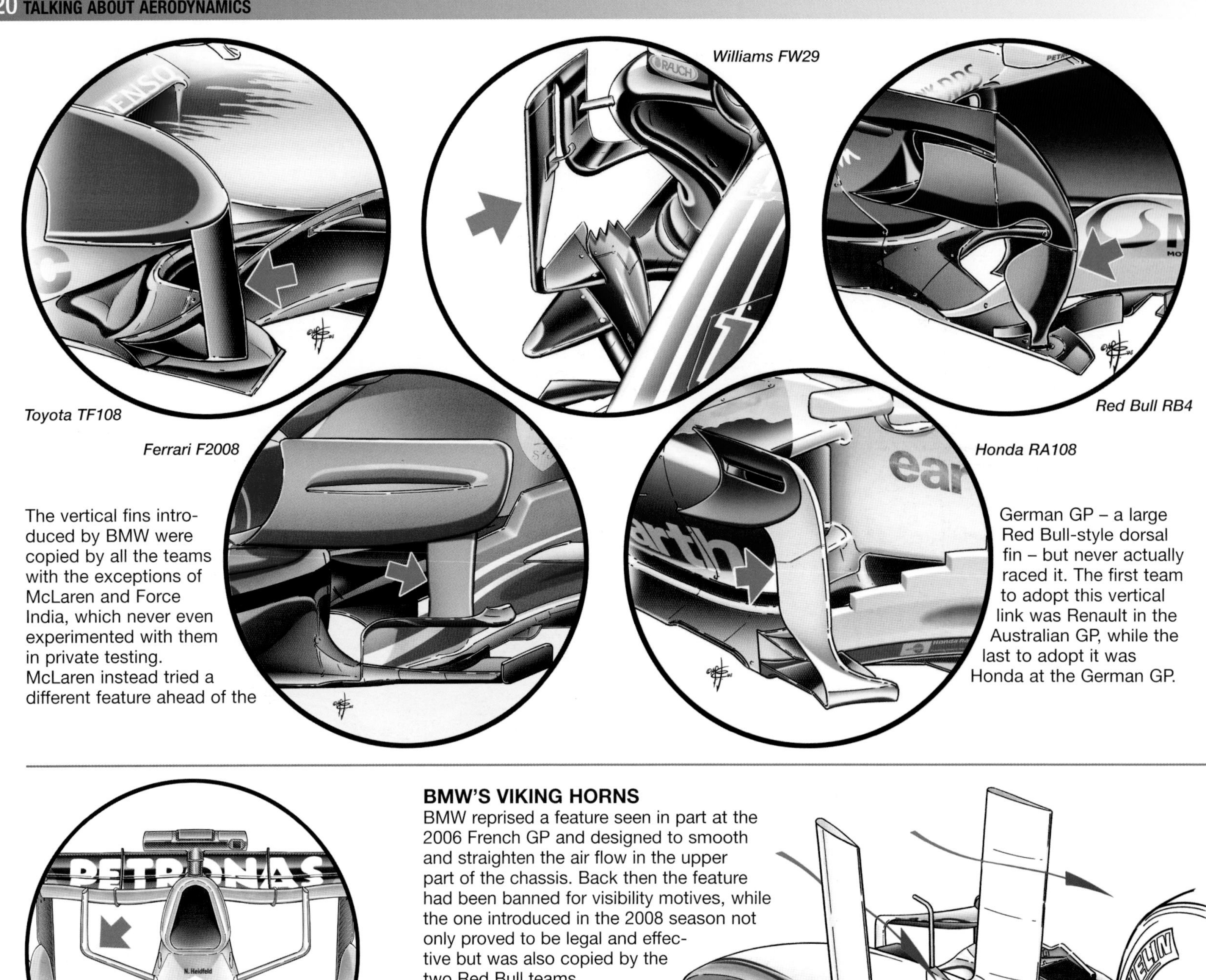

The vertical fins introduced by BMW were copied by all the teams with the exceptions of McLaren and Force India, which never even experimented with them in private testing. McLaren instead tried a different feature ahead of the German GP – a large Red Bull-style dorsal fin – but never actually raced it. The first team to adopt this vertical link was Renault in the Australian GP, while the last to adopt it was Honda at the German GP.

BMW'S VIKING HORNS

BMW reprised a feature seen in part at the 2006 French GP and designed to smooth and straighten the air flow in the upper part of the chassis. Back then the feature had been banned for visibility motives, while the one introduced in the 2008 season not only proved to be legal and effective but was also copied by the two Red Bull teams.
The feature also references the Viking horns introduced by McLaren in 2005 in the area of the dynamic engine air intake, later abandoned by the Woking team at the 2007 German GP but still used by BMW.

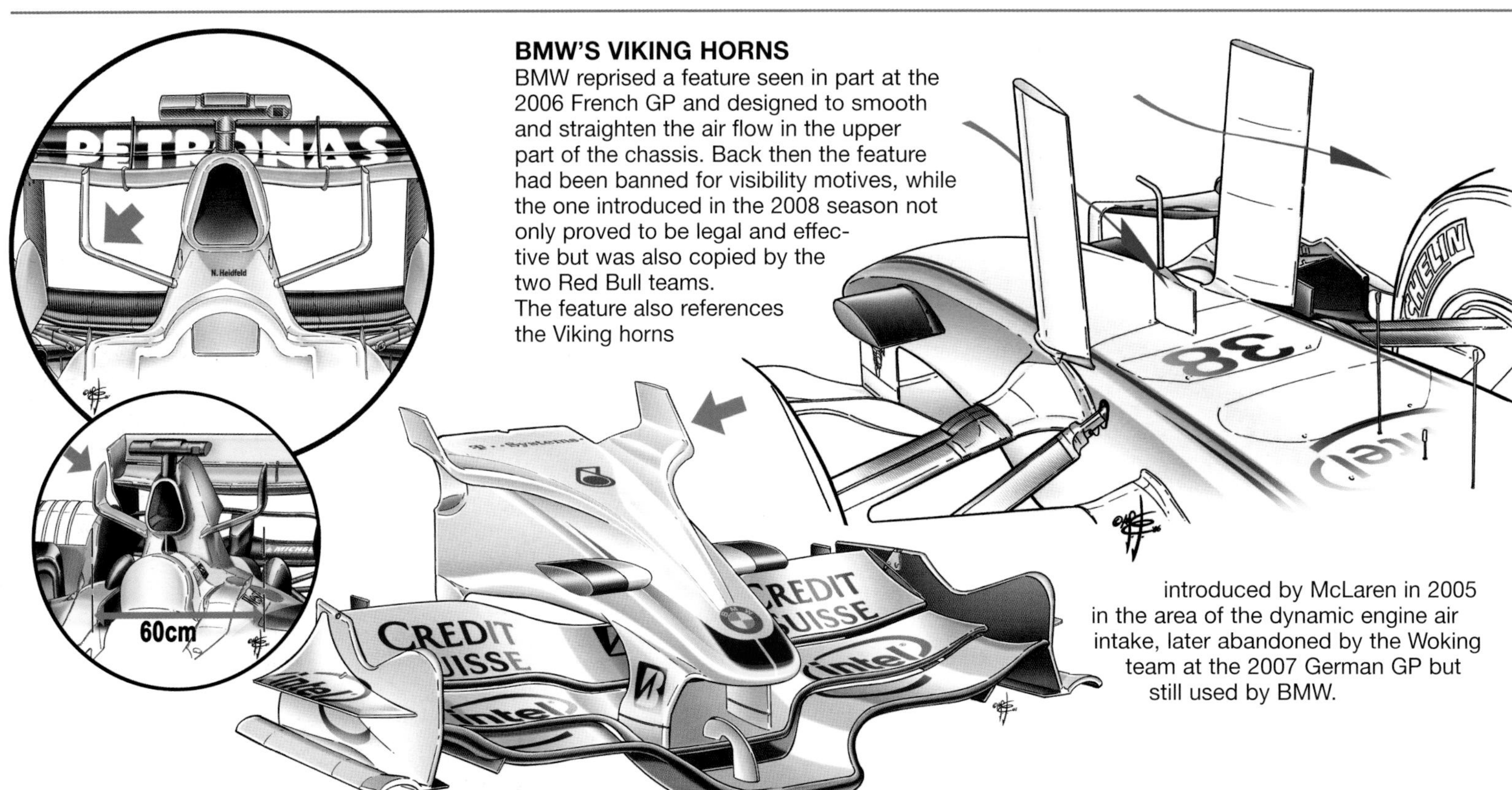

ELEPHANT'S (RABBIT'S) EARS

This feature had already been introduced by Honda in testing prior to the 2007 season but was never taken to the GPs. It was reprised from the 2008 Turkish GP onwards and then copied and employed on a number of occasions by McLaren too. In the case of the Honda its use required the car's nose to be lowered if it was to be effective which meant that a new crash test had to be passed. The aim of the feature was principally that of aligning the frontal air flows and directing them towards the rear of the car – cockpit, sidepods, engine air intake – but also to provide a slight increase in front end loading.The version used by McLaren, which already had a lower nose than the original Honda, was practically identical, the "ears" simply being attached. BMW introduced a feature that was replicated in the area either side of the engine cover. In the front view of the car you can see how the profiles are virtually aligned, almost as if to form an ideal channel for the air directed towards the rear wing.

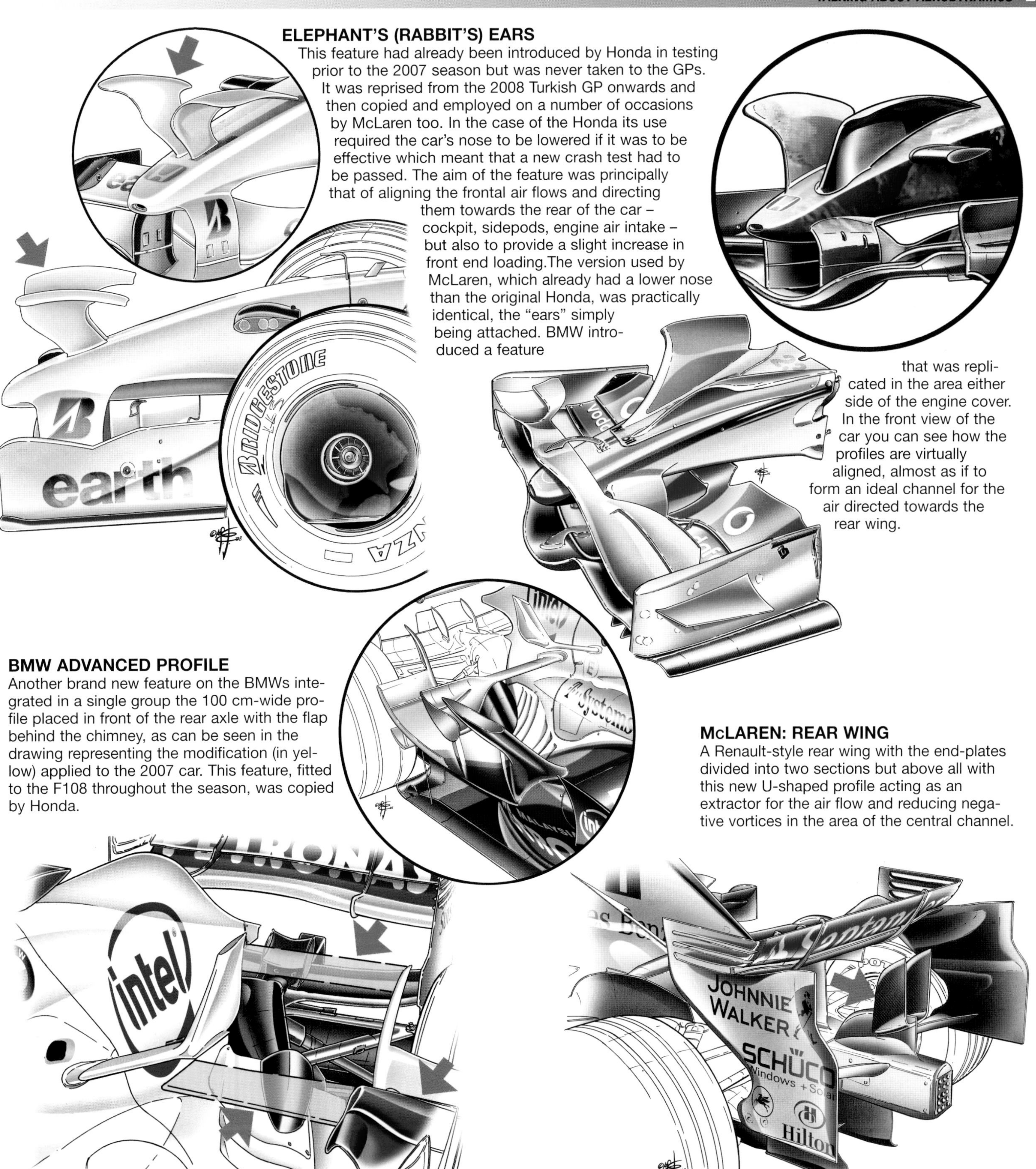

BMW ADVANCED PROFILE

Another brand new feature on the BMWs integrated in a single group the 100 cm-wide profile placed in front of the rear axle with the flap behind the chimney, as can be seen in the drawing representing the modification (in yellow) applied to the 2007 car. This feature, fitted to the F108 throughout the season, was copied by Honda.

McLAREN: REAR WING

A Renault-style rear wing with the end-plates divided into two sections but above all with this new U-shaped profile acting as an extractor for the air flow and reducing negative vortices in the area of the central channel.

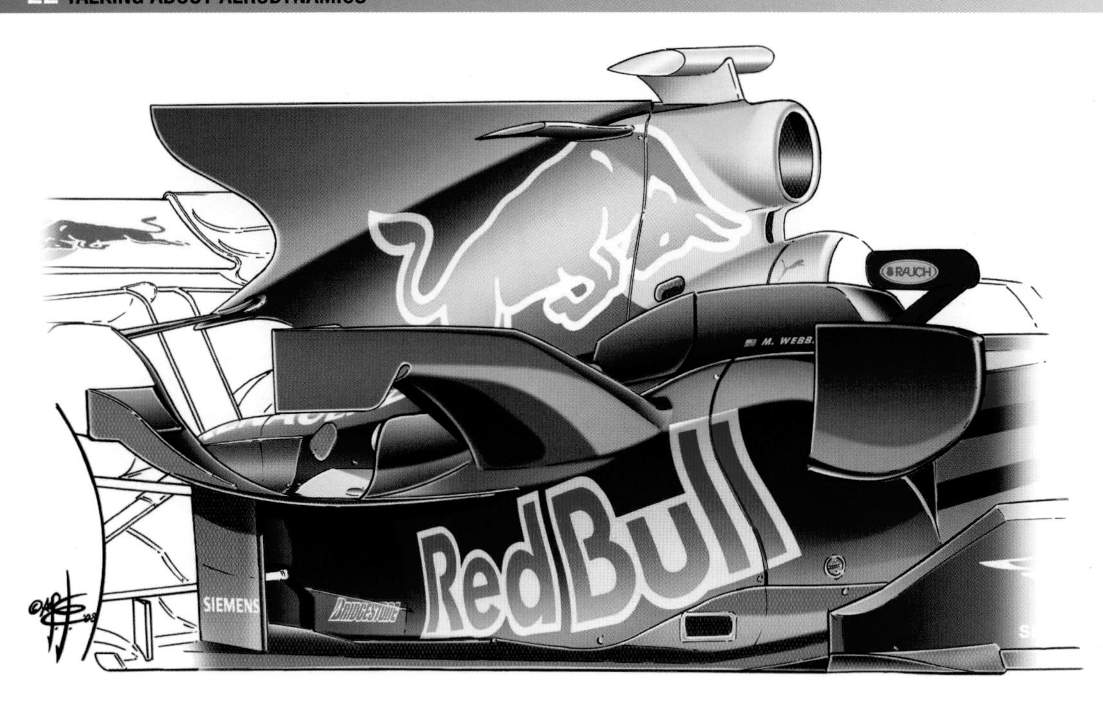

RED BULL FIN

This was the new feature with the longest history, back as 1972 with the Ferrari B2 and another in 1995 with the McLaren MP4-10 . The extension of the rear part of the engine cover was tapered and flattened.
This was done with the aim of providing greater stability when the car was yawing; that is to say, through corners, especially at high speed.
Aerodynamic penetration under these conditions is slightly inferior as a greater section is exposed to the air flow.
Furthermore, the rear wing receives a disturbed flow and is partially obscured by the presence of the lateral vane.
The fin, however, acts like that of an aeroplane or the sail of a boat; the air pushes on its surface and the flow is deviated parallel to the longitudinal axis of the car, increasing the efficiency of the profile and therefore the aerodynamic load.

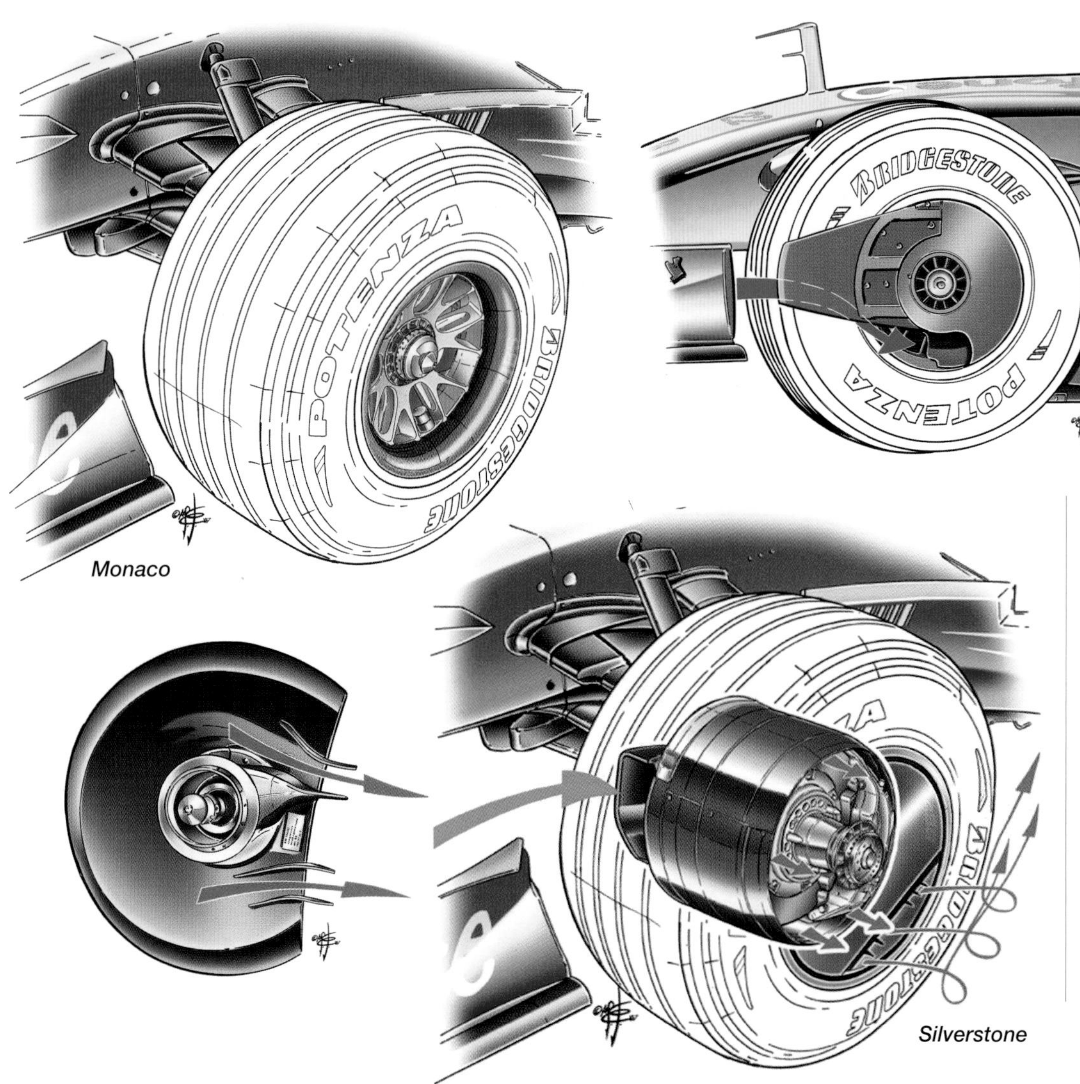

FRONT AND REAR WHEEL COVERS

In the 2007 season just Toyota and Williams had copied the front wheel covers introduced by Ferrari at Silverstone with the aim of better controlling the flow of hot air from the wheels. In 2008 virtually all the teams adopted the feature. On the Ferrari there was a sub-horizontal Gurney flap with the dual aim of deviating the flow of turbulent air caused by the steered wheels downwards and towards the outside of the car, but also to facilitate the extraction of hot air from the wheel hub-brake disc assembly. McLaren tested but never raced a feature that comprised a forward extension designed to reduce the turbulence generated by the front wheels.
The air vent on the Ferrari, McLaren and

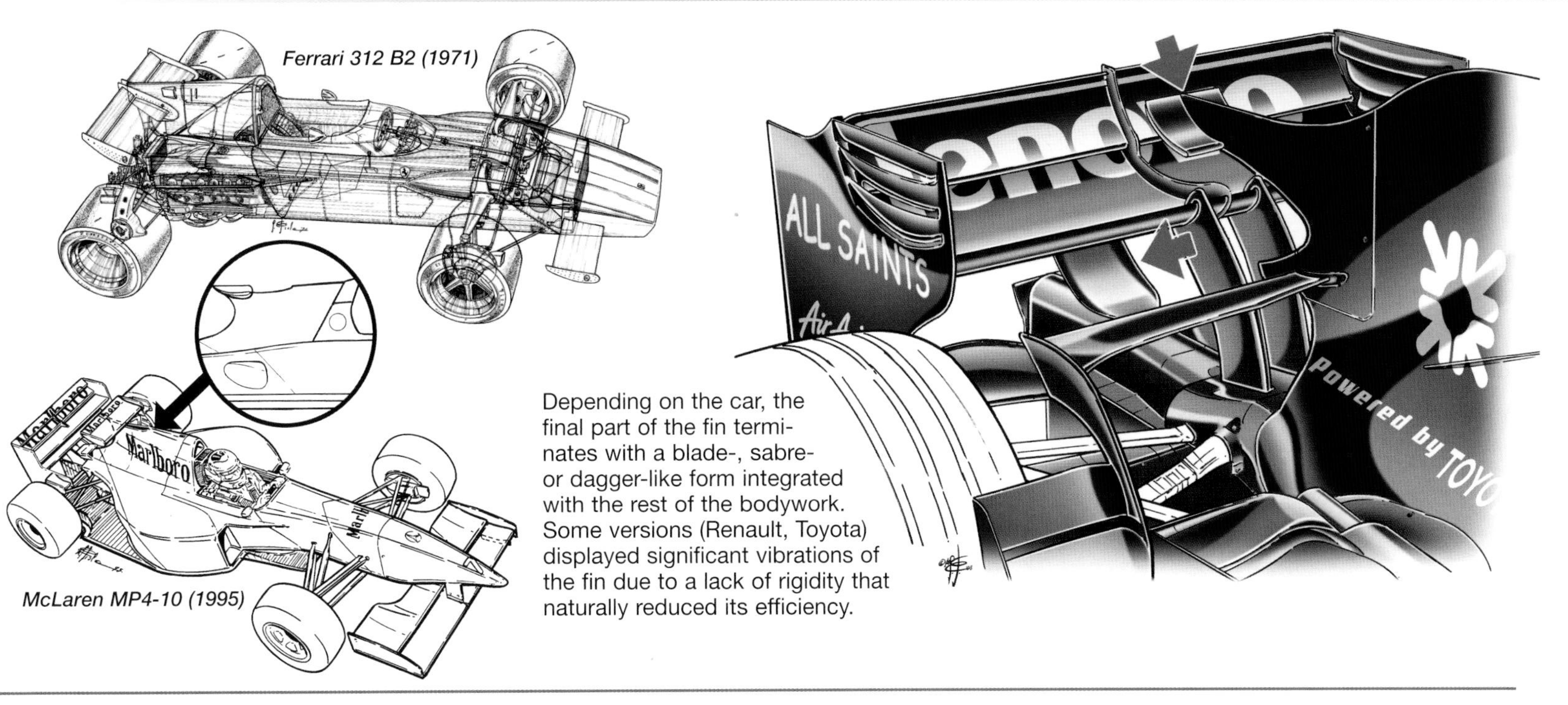

Ferrari 312 B2 (1971)

McLaren MP4-10 (1995)

Depending on the car, the final part of the fin terminates with a blade-, sabre- or dagger-like form integrated with the rest of the bodywork. Some versions (Renault, Toyota) displayed significant vibrations of the fin due to a lack of rigidity that naturally reduced its efficiency.

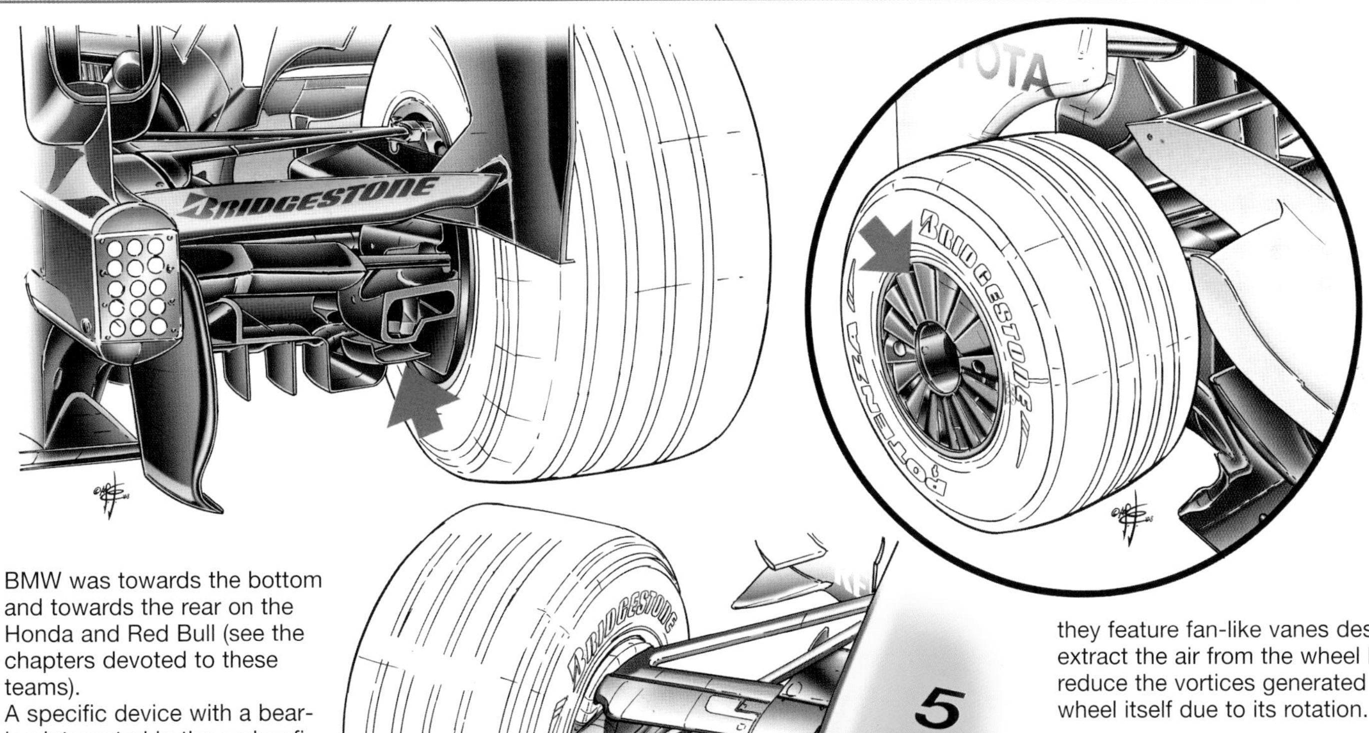

BMW was towards the bottom and towards the rear on the Honda and Red Bull (see the chapters devoted to these teams).
A specific device with a bearing integrated in the carbonfibre fairing and a rotation lock allows the wheel cover once mounted to remain in position and not rotate together with the wheel.
At the rear, the wheel covers were integrated with the wheel. Toyota presented an interesting variation on the rear wheels: they feature fan-like vanes designed to extract the air from the wheel hub rather than reduce the vortices generated around the wheel itself due to its rotation.

Brake cooling air intakes increasingly became true aerodynamic devices. Those at the rear had a pure downforce role with diverse winglets, while at the front their principal objective was that of shielding the internal channel between wheel and chassis from the turbulence generated by the wheels. The introduction of this mini-vane by Renault was innovative but was not a feature copied by other teams.

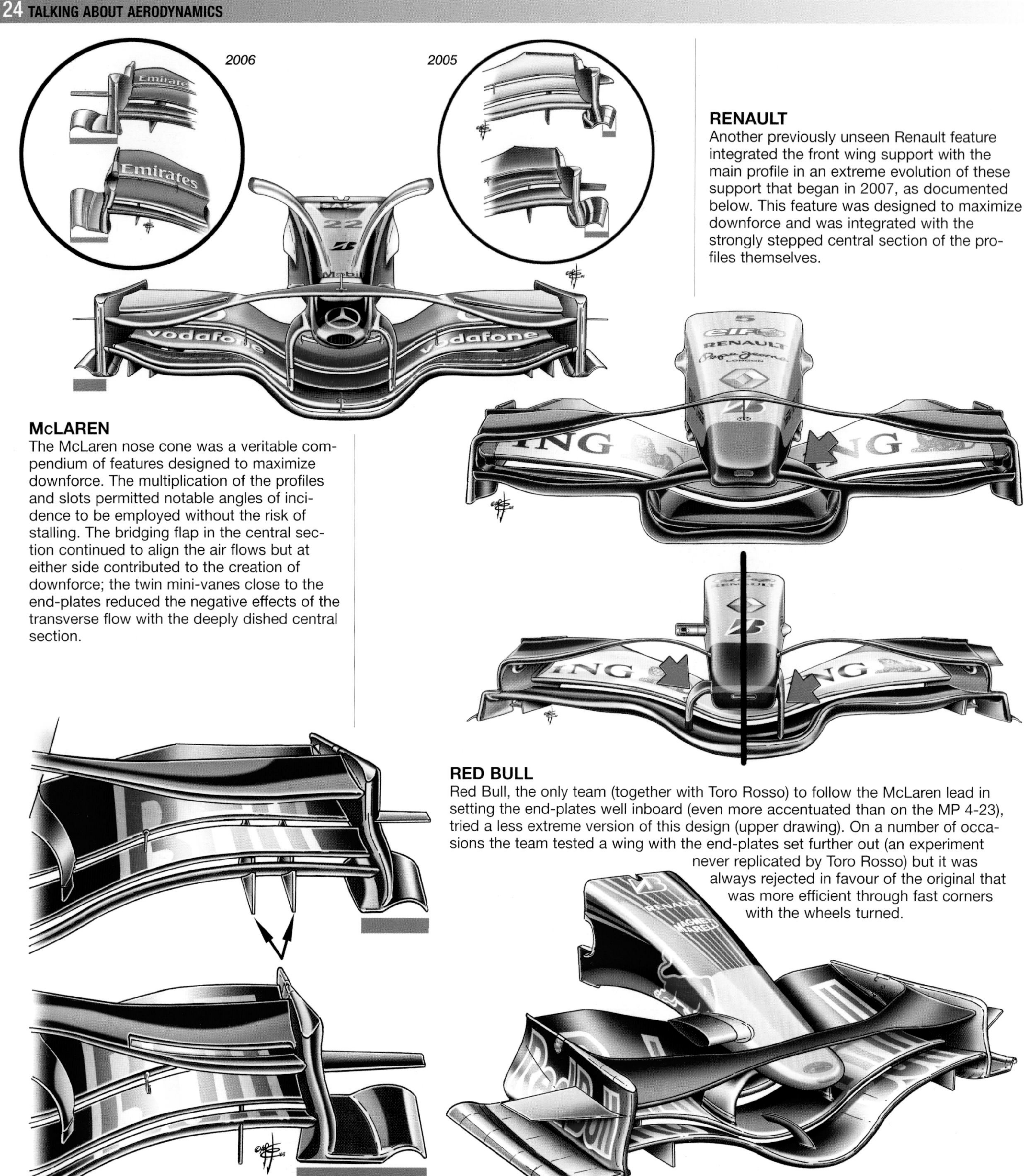

RENAULT

Another previously unseen Renault feature integrated the front wing support with the main profile in an extreme evolution of these support that began in 2007, as documented below. This feature was designed to maximize downforce and was integrated with the strongly stepped central section of the profiles themselves.

McLAREN

The McLaren nose cone was a veritable compendium of features designed to maximize downforce. The multiplication of the profiles and slots permitted notable angles of incidence to be employed without the risk of stalling. The bridging flap in the central section continued to align the air flows but at either side contributed to the creation of downforce; the twin mini-vanes close to the end-plates reduced the negative effects of the transverse flow with the deeply dished central section.

RED BULL

Red Bull, the only team (together with Toro Rosso) to follow the McLaren lead in setting the end-plates well inboard (even more accentuated than on the MP 4-23), tried a less extreme version of this design (upper drawing). On a number of occasions the team tested a wing with the end-plates set further out (an experiment never replicated by Toro Rosso) but it was always rejected in favour of the original that was more efficient through fast corners with the wheels turned.

In the last season of rules stability in the field of aerodynamics it was possible to register with a reasonable degree of accuracy the ability of each individual component of a car to create downforce, revealing values that the F1 legislators intended to modify drastically on the basis of the restrictions to be introduced for the start of the 2009 season (see the 2009 Season chapter). As can be seen from the data, the lower bodywork of the car plays the most important role in terms of aerodynamics, while the front and rear wings are above all designed to balance the car to best effect on the basis of the characteristics of the various circuits. It should however be noted that every component interacts with the others, improving their effectiveness, and for this reason it is difficult to quantify the true value of each individual piece. The division into four principal sectors is designed to give at least an idea of their general influence.

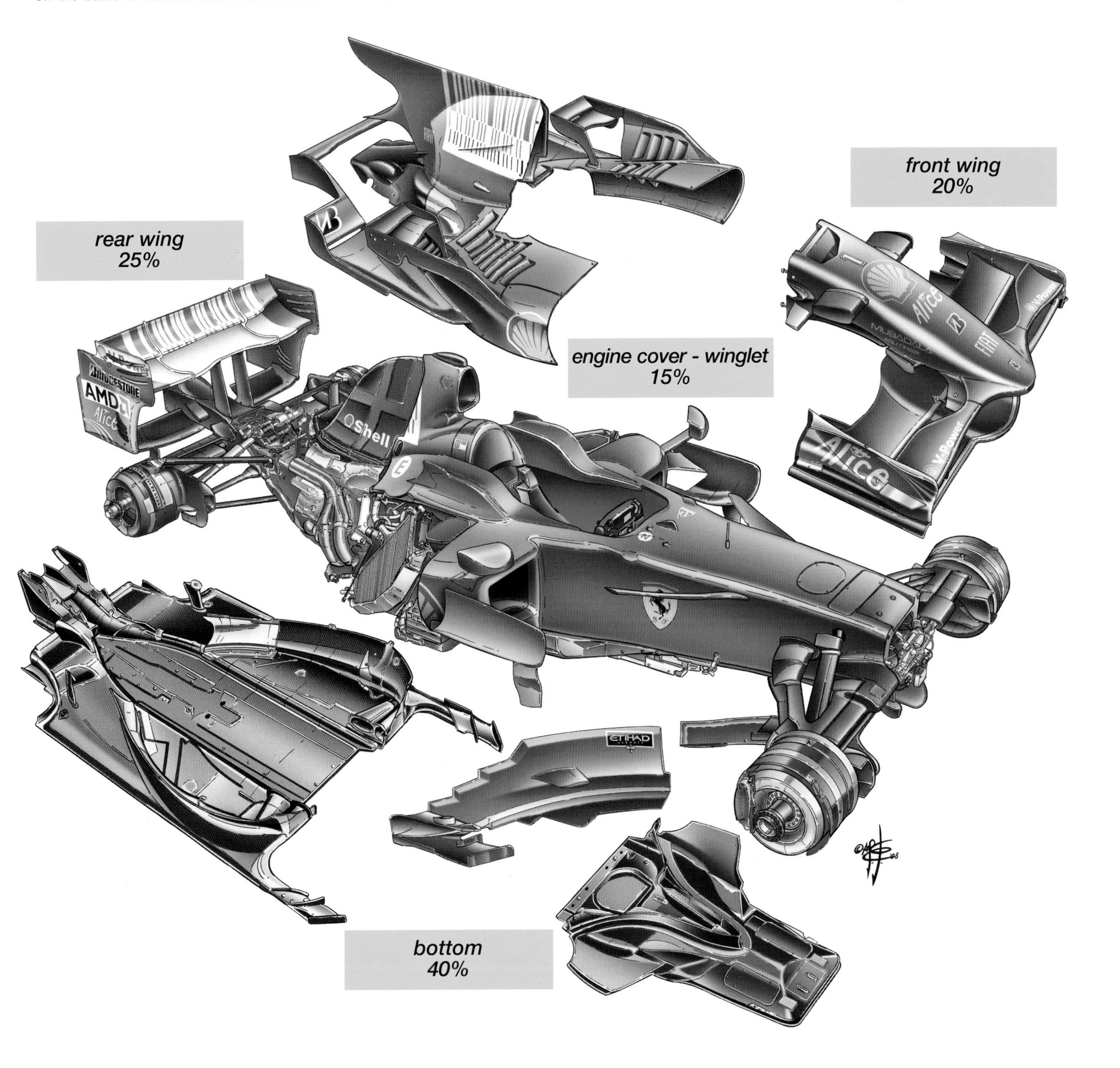

Bridgestone continued to supply the field with a range of identical tyres, 2008 the first year the formula was by regulation single-supply. In the previous year it had been so by default, following the withdrawal of Michelin. This was the 11th and final season for grooved tyres, prior to the re-introduction of slicks for 2009.
In terms of compounds, the soft was moved up the scale a little, made slightly harder, to make a better global fit with all of the circuits. In terms of construction the front shoulder was strengthened following Hamilton's Turkey '07 blow-out. Otherwise the tyres were as before, something in between a crossply and radial in construction, with a very rigid sidewall, something that led to a certain reluctance to generate heat in the carcass. Getting the necessary heat into the tyres for a qualifying lap and combining it with a set up that did not induce too much heat over a race stint was very much a key endeavour of the teams. Weight distribution, aero balance, suspension kinematics and driving styles were all crucial factors in the control of the teams in pulling off this delicate balancing act. Track temperature and the degree to which its surface rubbered in were the other important factors that were outside their control.
When a tyre compound is too soft for the demands of the track, it tends to grain. The soft rubber is simply not strong enough to support the grip and tears, forming grains of rubber on the surface. In these circumstances a harder compound will be faster. A tyre is more prone to graining when it is below its ideal operating temperature, as the rubber becomes more brittle and easily torn.
It is frequently the case that the softer tyre is quicker over one lap of qualifying - because the lap can be completed before the graining has become excessive – but slower over a race stint. As a track surface gets a layer of rubber upon it however – or the track surface temperature increases – the softer tyre will then tend to come into its working range and become faster than the harder. Its softer compound will chemically bond with the surface more effectively, increasing the grip and thereby the carcass temperature too. But occasionally the track surface can evolve yet further – as happened in Bahrain – inducing such grip from the softer tyre that it overheats its carcass, reducing its mechanical grip and in such a situation the harder tyre will again be the quicker one. It was this that allowed the few cars on the harder tyre at the end of the race in Bahrain – the BMWs, Kovalainen's McLaren and Button's Honda – to be unrepresentatively fast.
The Ferrari F2008 was generally the easiest on its tyres of all the cars, as befitting the team's long experience of Bridgestone. The McLaren was less aggressive on the rubber than in 2007 but still harder than the Ferrari. This occasionally compromised Ferrari's qualifying efforts, their drivers sometimes beginning their lap with the fronts still not up to the necessary 80deg C minimum. But this happened less frequently than in the past and the payback was usually better, more consistent tyre performance through a stint.
The combination of the McLaren's characteristics with Lewis Hamilton's aggressive driving style ensured that Istanbul's turn eight still presented a problem for Hamilton's right-front, despite the strengthening of the shoulder. The problem was simply manifest in the sidewall instead and the Bridgestone engineers calculated that the safe maximum for Hamilton in the race was 15 laps before the tyres would need to be changed. This enforced a less than optimum three-stop strategy upon him. This is the highest-duration and highest lateral-g corner on the F1 calendar.
Hamilton would also occasionally find that he could not access the theoretical extra grip of the softer compound on any given race weekend because he had a tendency to overheat the rears, thereby making the harder tyre the faster one for him. This happened in both Istanbul and Barcelona.
The biggest downside of the Ferrari's gentle usage characteristic came in wet weather on dry-weather tyres. Whereas the challenge in a dry race is to prevent the rubber overheating, in the wet it is the opposite: keeping enough

	Bridgestone		McLaren							
	Prime	Option	Hamilton				Kovalainen			
Stint			1	2	3	4	1	2	3	4
Australia	medium ●	soft □	●	●	□		●	●	□	
Malaysia	hard ▲	medium ○	○	○	▲		○	▲	○	
Bahrain	medium ●	soft □	□	□	●		□	□	●	
Spain	hard ▲	medium ○	○	○	▲		○			
Turkye	hard ▲	medium ○	▲	▲	▲	○	▲	▲	▲	○
Monaco	soft ■	super soft ◇	★	★	■		★	★	■	
Canada	soft ■	super soft ◇	◇	◇			■	◇	◇	
France	medium ●	soft □	□	●	□		●	●	□	
Great Britain	hard ▲	medium ○	★	★	★		★	★	★	
Germany	hard ▲	medium ○	▲	▲	○		▲	▲	○	
Hungary	soft ■	super soft ◇	■	■	◇		■	■	◇	
Europe	soft ■	super soft ◇	■	■	◇		■	■	◇	
Belgium	hard ▲	medium ○	○	○	▲		○	○	▲	
Italy	hard ▲	medium ○	❊	❊	★		❊	❊	★	
Singapore	soft ■	super soft ◇	■	■	◇		■	◇		
Japan	medium ●	soft □	●	●	□		●			
China	hard ▲	medium ○	▲	▲	○		▲	▲	○	
Brazil	medium ●	soft □	★	●	●	★	★	●	●	★

	Renault							
	Alonso				Piquet			
Stint	1	2	3	4	1	2	3	4
Australia	●	●	□		●	●		
Malaysia	○	▲	○		○	○	▲	
Bahrain	□	□	●		●	□	□	
Spain	○	○			○			
Turkye	▲	▲	○		▲	▲	○	
Monaco	★	❊	❊	■	❊	■		
Canada	■	■			■	■		
France	●	●	□		●	●	□	
Great Britain	★	★	★		★	★		
Germany	▲	○	▲		▲	○		
Hungary	■	■	◇		■	■	◇	
Europe	■				■	◇		
Belgium	○	○	▲	★	▲			
Italy	❊	★			❊	★		
Singapore	◇	■	■		■			
Japan	●	●	□		●	●	□	
China	▲	▲	○		▲	▲	○	
Brazil	★	●	●	★	★			

	Ferrari								
	Raikkonen					Massa			
Stint	1	2	3	4	5	1	2	3	4
Australia	□	●				□	□	□	
Malaysia	▲	▲	○			▲	▲		
Bahrain	□	□	●			□	□	●	
Spain	○	○	▲			○	○	▲	
Turkye	○	○	▲			○	○	▲	
Monaco	★	★	★	◇	◇	★	★	◇	
Canada	■	■				■	■	■	◇
France	●	●	□			●	●	□	
Great Britain	★	★	★			★	★	★	★
Germany	▲	○	○			▲	○	▲	
Hungary	■	■	◇			■	■	◇	
Europe	■	■	◇			■	■	◇	
Belgium	○	○	▲			○	○	▲	
Italy	❊	❊	★			❊	❊	★	
Singapore	■	■	◇			■	■	◇	
Japan	●	●	□			□	□	●	
China	○	▲	○			○	○	▲	
Brazil	★	●	●	★		★	●	●	★

	Honda									
	Button					Barrichello				
Stint	1	2	3	4	5	1	2	3	4	5
Australia	●					●	●	□		
Malaysia	○	▲	○			▲	○	○		
Bahrain	□	□				□	□	●		
Spain	○	○	▲			○	○	○		
Turkye	▲	○				○	▲			
Monaco	★	★	★	■		★	◇			
Canada	■	◇	■	■		■	◇			
France	●	●				●	●	□		
Great Britain	★	★	❊			★	★	❊	★	
Germany	○	▲	○	○		○	○	▲		
Hungary	■	■	◇			■	■	◇		
Europe	■	◇				■	◇			
Belgium	○	▲	★			○	○			
Italy	❊	★	★			❊	★	○		
Singapore	■	■	◇			■	■			
Japan	□	●				●	□			
China	▲	▲	○			▲	▲	○		
Brazil	★	●	●	●	❊	★	●	●	❊	★

	BMW Sauber							
	Heidfeld				Kubica			
Stint	1	2	3	4	1	2	3	4
Australia	●	●	□		●	●	□	
Malaysia	○	▲	○		▲	▲	○	
Bahrain	□	□	●		□	□	●	
Spain	○	○	▲		○	○	▲	
Turkye	▲	▲	○		▲	▲	○	
Monaco	★	★	■	■	★	★	■	
Canada	■	◇			■	■	◇	
France	●	●	□		●	●	□	
Great Britain	★	★	★		★	★	★	
Germany	○	○	▲		○	○	▲	
Hungary	■	◇			■	■	◇	
Europe	■	■	◇		■	■	◇	
Belgium	○	○	▲	★	○	○	▲	
Italy	❊	★			❊	★		
Singapore	■	■	◇		■	■	◇	
Japan	●	□			●	●	□	
China	▲	▲	○		▲	▲	○	
Brazil	★	●	●	★	★	●	●	★

wet ★ extra wet ❊

Talking about TYRES and BRAKES

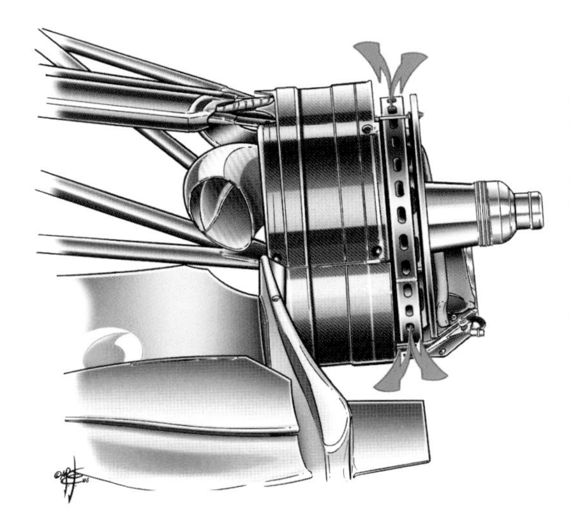

At the Belgian GP McLaren came up with a strategy to bring the front tyres up to temperature more quickly despite the low environmental temperatures and rain. The team partially removed the covers that usually enclose the discs and calipers and channel the hot air towards the outside of the wheel in order to keep it within the wheel instead. In this way the temperature of the tyres increased, guaranteeing enhanced grip.

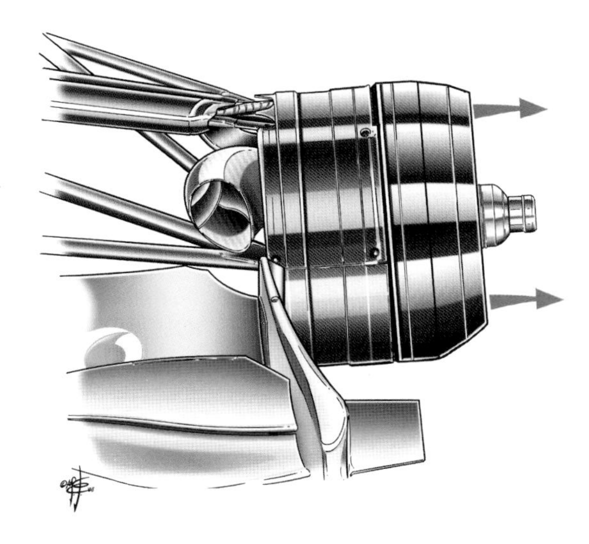

heat in the tyres to prevent the chemical bonding with the surface from breaking down is essential. The Ferrari was generally unable to retain tyre heat when the rain came whereas the McLaren was especially effective. This was seen to dramatic effect in the late stages of the Belgian Grand Prix where Raikkonen's race-leading Ferrari was suddenly devoid of grip as rain arrived in the last three laps, allowing Lewis Hamilton's McLaren an enormous grip advantage.

This characteristic played its part in determining the outcome of the championship at the final dramatic race in Brazil.

When rain arrived just a few laps from the end, most teams opted to pit their drivers for a change to intermediate tyres. But Toyota chose to leave their men Timo Glock and Jarno Trulli out on their dry weather tyres, hoping they would retain their heat long enough to make use of the time saved by not pitting. It worked, in the sense that it moved Glock up from eighth to a sixth place finish, but the catastrophically sudden reduction in grip on the final lap as the tyres suddenly fell out of their operating temperature range allowed the inters-shod Hamilton to overtake him, thereby gaining enough points to take the title.

It emphasised to the wider world what racing engineers have always known: that tyres are the primary performance factor.

Mark Huges

Toyota								Red Bull							
Trulli				Glock				Coulthard				Webber			
1	2	3	4	1	2	3	4	1	2	3	4	1	2	3	4
●				□	●			●	●			●			
○	○			○				○	○	▲		▲	▲	○	
□	□	●		□	□	●		□	□	□	●	□	□	●	
○	○	▲		○	○	▲	▲	○	○	▲	○	○	○	▲	
▲	○	○		○	▲			▲	▲	○		▲	▲	○	
★	✻	★	◇	★	✻	★	◇	★				★	■		
◇	■			◇	■			■	◇			■	■	◇	
●	●	□		●	●	□		●	●	□		●	●	□	
★	★	★		★	★	★		★				★	★	★	
○	○	▲		○	○			○	○	▲		○	○	▲	
■	■	◇		■	■	◇		◇	■	■		■	■	◇	
■	■	◇		■	◇			■	■	◇		■	◇		
○	○	▲		○	▲	★		○	▲	★		○	○	▲	
✻	✻	★		✻	✻	★		✻	★	★		✻	✻	★	
■	◇			■	■	◇		■	■	◇		■	■		
●	●	□		●	●	□		●				●	□		
▲	▲			▲	○			○	▲			▲	○	○	
★	●	●		★	●	●		★				★	●	●	★

Williams									Force India							
Rosberg					Nakajima				Fisichella				Sutil			
1	2	3	4	5	1	2	3	4	1	2	3	4	1	2	3	4
●	●	□			●	●	□		●				□			
○	○	▲			○	○	▲		○	▲	▲		▲			
●	□	□			●	□			□	□	●		□	□	□	●
○	○				○	○	▲		○	○	▲		○			
▲	▲	○			▲				▲				○	▲	▲	▲
★	★	★	★	■	★	■			★				★	■		
◇	◇	■	◇		■	■			■	■	◇		◇			
□	●				□	●	●		●	●	□		□	●	●	
★	★	✻	★		★	★	✻		★				★			
○	○	▲			○	○	▲		▲	○	○		▲	○	○	
■	■	◇			■	◇			■	■	◇		■	◇	■	■
◇	■	■			■	◇	◇		■	◇			■	◇	◇	
○	▲	✻			○	▲	✻		○	○	▲	★	○	○	▲	
✻	✻	★			✻	★			✻				✻	✻	★	○
◇	■	◇			■	■	◇		■	◇			■	■	◇	
●	□				●	□			●				●			
○	○	▲			○	▲			▲	○			○			
★	●	●	★		★	●	●	★	★	□	□	✻	★	□	□	✻

Toro Rosso									Super Aguri							
Vettel					Bourdais				Sato				Davidson			
1	2	3	4	5	1	2	3	4	1	2	3	4	1	2	3	4
●					●	□	●		●	●			●			
○	○				○				○	○	▲		○	○	▲	
●					●	□	□		□	□	●		□	□	●	
○					○				○	○	▲		○			
▲	▲	▲	○		▲	▲										
★	■				★											
■	◇				■	◇	■									
●	●	□			●	●	□									
★					★	★	✻									
○	○	▲			▲	○	▲									
■	◇				■	■	◇									
■	■	◇			■	■	◇									
○	○	▲			○	○	▲									
✻	✻	★			✻	✻	★									
■	■	◇			■	◇	■									
●	□				●	○										
▲	▲	○			▲	○	○									
★	●	●	●	★	★	●	●	✻								

The development of braking systems begun in 2007 continued the following year, further emphasising their importance relative to the car's aerodynamics.
In the design of the car, the complexity of the corner (the upright-caliper together) is taken to the extreme with the brake air intakes that have become increasingly integrated into the aerodynamic elements inside the wheel (drums, doughnut etc) to the point of now "covering" the brake disc in every direction. The principal objective is to convey and direct the air flow, which is then extracted through the flanges on the rims for the purpose of improving downforce and reducing aerodynamic resistance, as well as to provide the amount of air necessary to dissipate the heat generated during braking. It is in that sense that the almost total generalisation of "wind shields" or rim covers on the front wheels – introduced by Ferrari for the 2007 GP of Great Britain – should be seen, with the obvious task of interacting with the air flow in the delicate area of the front wheels.
From this point of view, both the ventilation of the discs and the brake calipers are designed not just to minimise operating temperatures, but also and above all so as not to counter the aerodynamic effects already described.
That complexity translates into a different design choice for each team in relation to the various chassis and aerodynamic configurations, starting with the installation of the caliper on the upright, which must satisfy both the fixed bulk of the suspension's geometry and matters related to weight distribution and cooling.

Ferrari

The supply of the braking systems for the 2008 season saw an advancement in the presence of Brembo, which from the start of the season provided its systems to BMW Sauber, Ferrari, Red Bull, Scuderia Toro Rosso and Toyota and which, from the race at Spa, also equipped the Hondas when they abandoned their exclusive supply by Alcon. The others were Britain's A-P for Renault, Williams and Force India, while McLaren continued its exclusive experience with Japan's Akebono. The most demanding challenge of 2008 was the management of the design, testing and development of six different systems for the bulk, thickness and geometry of the discs and pads, cooling and the ratio of weight/rigidity. Also bear in mind that the front and rear braking systems are often made up of different calipers for the two axles, for which reason the level of complexity doubles.
The braking system is increasingly becoming one of the initial elements in the conception of a new car for the implications that it has in the definition of all the aspects that make up a corner.

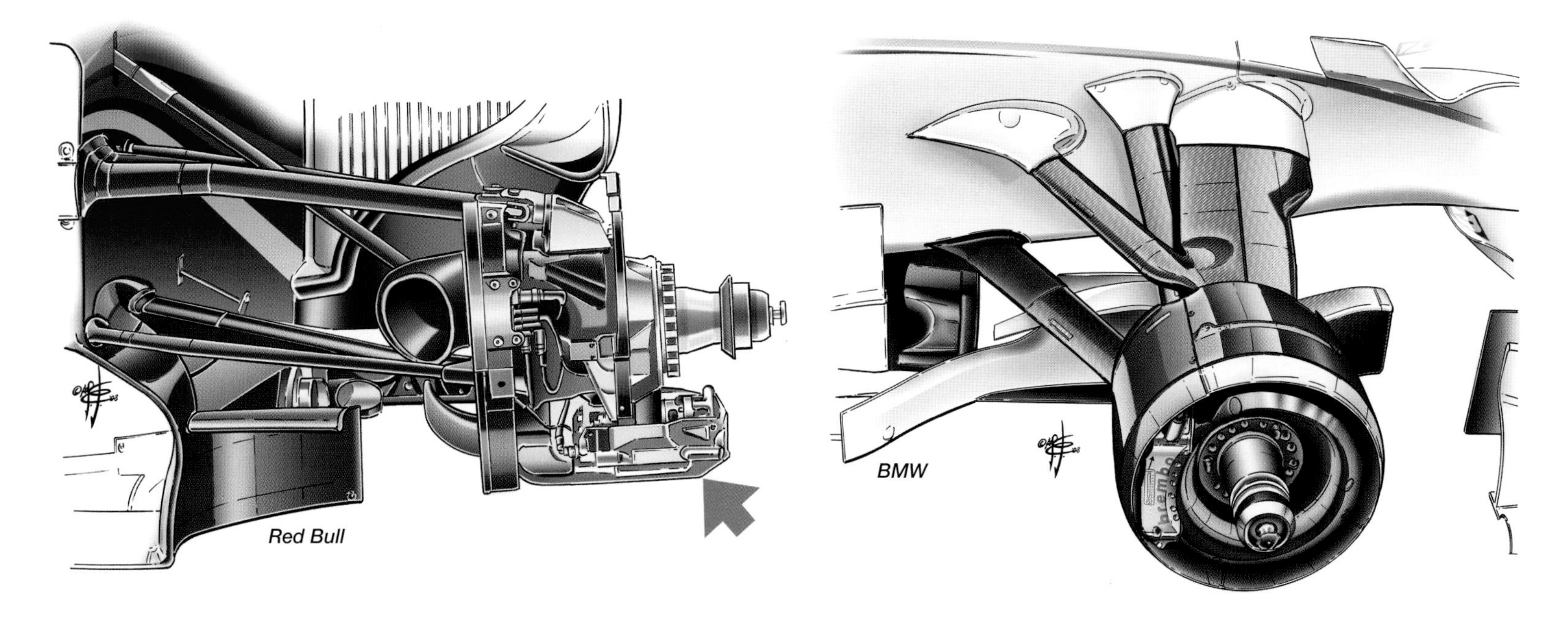

Red Bull

BMW

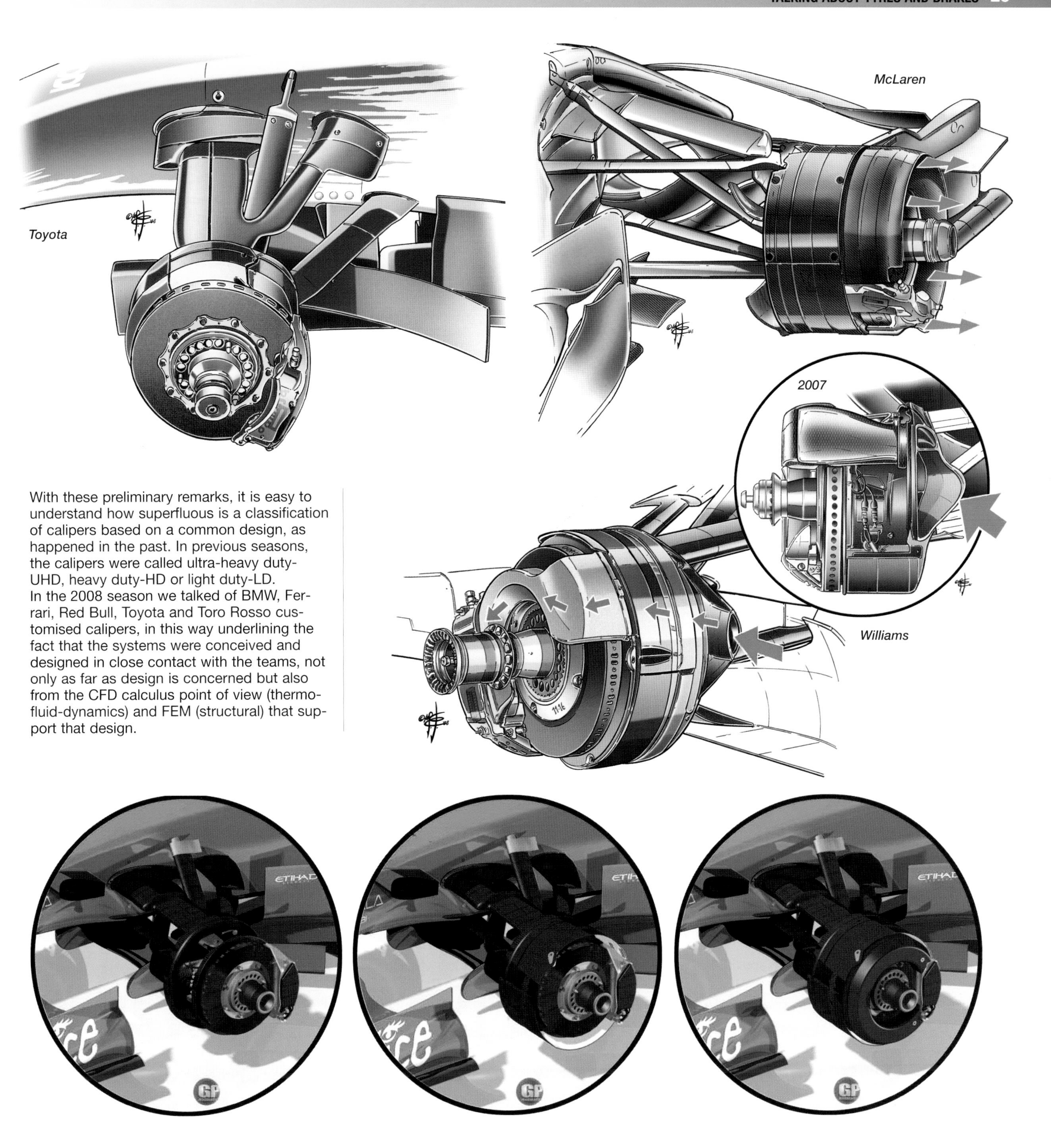

With these preliminary remarks, it is easy to understand how superfluous is a classification of calipers based on a common design, as happened in the past. In previous seasons, the calipers were called ultra-heavy duty-UHD, heavy duty-HD or light duty-LD.
In the 2008 season we talked of BMW, Ferrari, Red Bull, Toyota and Toro Rosso customised calipers, in this way underlining the fact that the systems were conceived and designed in close contact with the teams, not only as far as design is concerned but also from the CFD calculus point of view (thermo-fluid-dynamics) and FEM (structural) that support that design.

With the ban of many electronic aids it is reasonable to have thought that the 2008 steering wheels would be less sophisticated than those of the 2007 season. But right from the early races it was clear that, in this sector, there would be a substantial escalation in the complexity of the controls from the moment that, precisely in relation to the new regulations, the management of many functions that were previously controlled electronically were now to be carried out manually by the driver.
While the Regulations chapter clearly illustrates the greater complexity of the Ferrari steering wheel compared to the 2007, here we focus on the specific new features, which were numerous and all concentrated on the steering wheel. Ferrari introduced a sophisticated start system that enabled them to get partially around the obstacle of the abolition of the traction control. From the very first Grands Prix it was noted how the two Ferrari drivers traversed the pit lane at slow speed to then simulate a sort of racing start once they had reached the traffic light. That is how it was discovered that Ferrari had developed a start system that was perfectly legal and always managed by the driver, but which enabled the cars to shoot away from the grid very well, as happened for example at the Grand Prix of Bahrain, where Massa 'burnt' poleman Kubica.
The steering wheel/computer of the F2008 went back to having double paddles for the management of the gears (Schumacher preferred a single element with a double function so that he could make the changes with a single hand if necessary, as shown in the 2006 Technical Analysis) and has obviously retained the double clutch levers. The latter were always adopted to enable the driver to avoid the engine cutting out after a spin due to a sudden operation of one of the two levers. With the abolition of the launch control, the management of the start went back to being totally controlled by the hands of the driver, who was also called on to improve the sensitivity during the disengagement of the clutch, a method that can permit perfect starts without too much wheel spin. The double control, therefore, carried out an important role: these are the reasons.
The special mapping of engine power, which

McLAREN STEERING WHEELS

McLaren retained the classic shape of its steering wheel, but introduced the new feature of six paddles at the rear of it. To the four levers behind the wheel of the MP4-22 were added another two supplementaries positioned (2) so that they were almost attached to the traditional units for the gearbox. The purpose was to operate both paddles with a single movement of the hand, setting in motion two functions, the gear change and the appropriate mapping of the engine power , but also a sort of manual traction control. There was no change to the paddles that controlled the management of the gears (1) or the two (3) bigger units down below for the clutch.

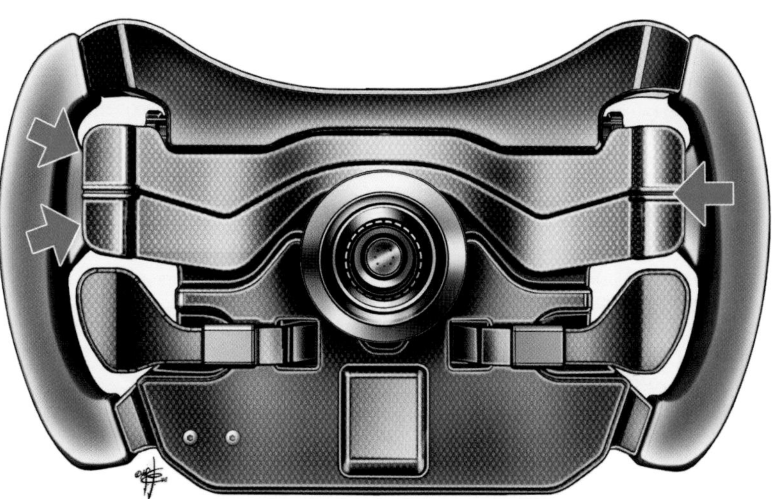

Talking about COCKPITS and STEERING WHEELS

by regulation must be maintained for at least 90 seconds, envisages at first a fairly flat early power curve with optimum revs of around 12,000 to then have a more brusque progression.
The McLaren answer in this race for steering wheel sophistication was to introduce yet another two paddles, taking the number to six of them behind the wheel, always in an attempt to make it more ergonomic, functional and 'targeted' on certain controls such as the management of the clutch or, in the case of McLaren, the management of various mappings of engine power, which integrate themselves with the action of changing gear, permit the recreation of a sort of manual traction control by the driver. A feature that was introduced from the first race of the season, as was pointed out by Martin Whitmarsh, managing director of McLaren, but which was discovered only after three races by the onboard camera. The new paddles were difficult to distinguish due both to their carbon colour and their extreme closeness with the traditional units of the gearbox.
As the season continued, another two teams followed McLaren's example: they were Renault and Honda, while Ferrari limited itself to just privately testing a six-paddle steering wheel, but afterwards decided to stay with the four paddle solution. Renault was the first team to follow McLaren's example, transforming the control that was usually managed by buttons on the wheel into a paddle. The Anglo-French team did so for the Grand Prix of Great Britain, but only on Piquet's car. After that, he was able to manage the differential with greater freedom without taking his hands off the wheel, while Alonso only began to use this modification two races later.
And finally, we note a new feature that could be considered a sort of 'legacy' by David Coulthard during his last season in Formula 1. It concerned safety inside the cockpit and was a kind of added belt that the Scottish driver had always had since his debut in F1 and which we have never been able to spotlight due to lack of space.
The curious fact is that during his 15-year career in F1 no other driver ever adopted this solution.

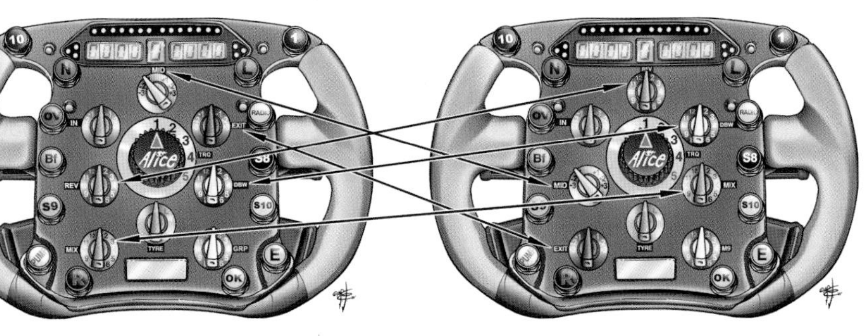

FERRARI
Completely identical of shape were the two steering wheels of the Ferrari drivers but they were personalised in the disposition of the various controls. It can be easily seen in the comparison with the crossed arrows how Massa (right) had three paddles for the differential (for entry, middle and exit from a corner) that were vertical and on the left hand side, while Raikkonen had them centred up high. This different disposition obviously had its influence on other controls, like the rev limiter (REV), the air/petrol mixture (MIX) and the management of the accelerator (DBW).

RENAULT

Renault was the first team to follow McLaren in adding 2 levers behind the steering wheel. It did so at the Grand Prix of Great Britain, two GPs after the discovery of the new feature on the MP4-22 at Monaco. The change was first adopted for the Piquet car, with two titanium paddles mounted up higher (1) and more separated from those of the gearbox (2) in relation to McLaren. The double paddles of the clutch were unchanged: they were needed by the Brazilian to alter the regulation of the differential and reduce the understeer that afflicted his car. From the Grand Prix of Hungary, Alonso also used this system. As you can easily see from the rear view, these paddles were added without changing the steering wheel.

FERRARI: DOUBLE CLUTCH CONTROL

To better manage the start in the absence of traction control, Maranello sub-divided the working of the two clutch levers: the first on the left (3) recalls the mapping of the electronic management system, which permitted at a certain number of engine revs to move the wheels without spin. Immediately afterwards, the driver was able to also let the second lever (4) go and only in that way was the full power let loose. The driver had to be able in know how to dose the movements, attempting to regulate his them to find the right synchronisation. And it was for this reason that the two Ferraristi acted in that way as they traversed the pit lane, also availing themselves of data collected by telemetry. The temperature of the tyres, the operating pressure, the type and condition of the asphalt and the weather conditions were variable so that elaboration by Magneti Marelli electronics enabled them to obtain the optimum track setting for the mapping to select the engine power (1) and the disengagement of the clutch (2).

HONDA

This team also went in the direction of adding two mini-paddles to the rear of their steering wheel to make it more ergonomic, a feature adopted first by Button and then by Barrichello.

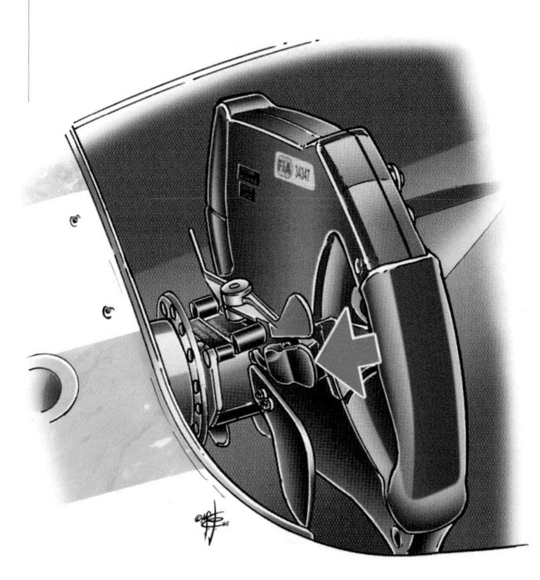

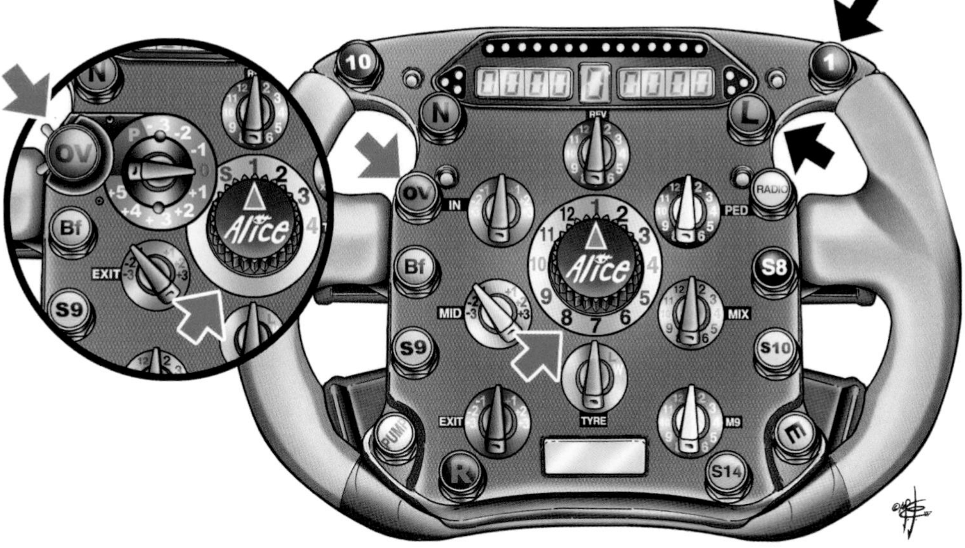

MASSA

Ferrari responded to the six paddles of the McLaren steering wheel with a further sophistication of their steering wheel of which only one example was available and was assigned to Felippe Massa from the Grand Prix of Hungary. A number of functions had been modified and moved to make it more ergonomic, like the central paddle (1), while the ability to operate at maximum revs was completely new, previously carried out using a simple button (2) and now by a paddle with various possibilities, dependent on the situation. On the start line, it was seen that the two clutch levers were of a different shape with a possible double use, as can be seen in the detail of the back of the steering wheel.

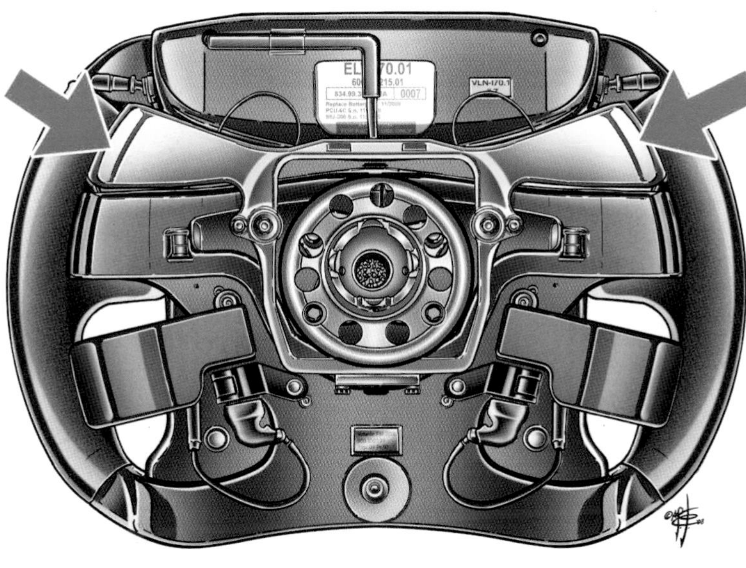

FERRARI'S SIX PADDLES

At the Monza tests in mid-August, Ferrari experimented with the six lever format, adding a rocker arm up high with a small titanium frame, which was easily visible from outside. But this steering wheel was never used during the days of the Grands Prix and remained in the experimental phase.

BMW

The steering wheels of the two BMW drivers were highly personalised, so much so that it made an exchange of steering wheels between the two impossible. Heidfeld preferred an anatomic grip in smooth leather, while Kubica had an almost normal rim covered in suede. The shape of the gearbox management paddles was also different, the German preferring his to be longer and higher.

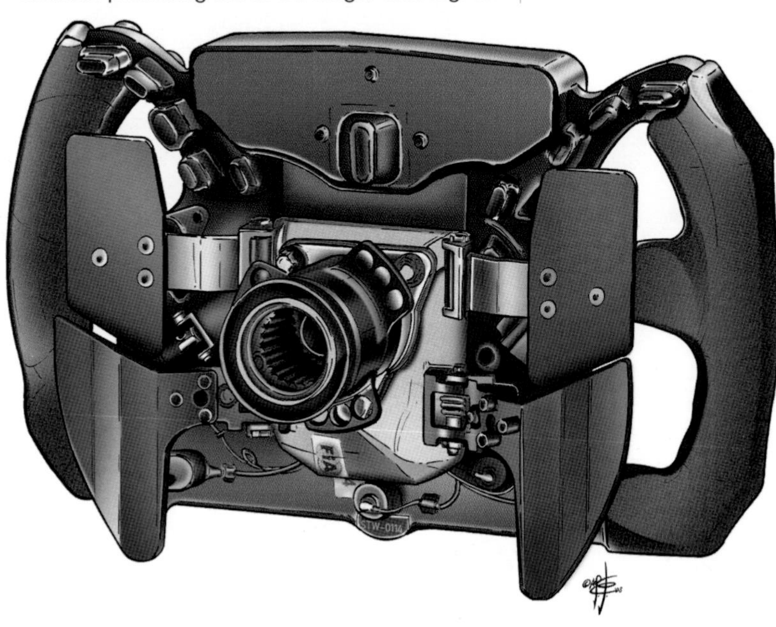

DAVID COULTHARD

From Williams to McLaren and lastly Red Bull, David Coulthard always used an additional belt to strap himself into the cockpit more effectively. It was a large, transverse strap that impeded the torso from moving under the effect of powerful lateral acceleration in fast corners. A modification destined to disappear with the definitive retirement of the Scotsman from Formula One.

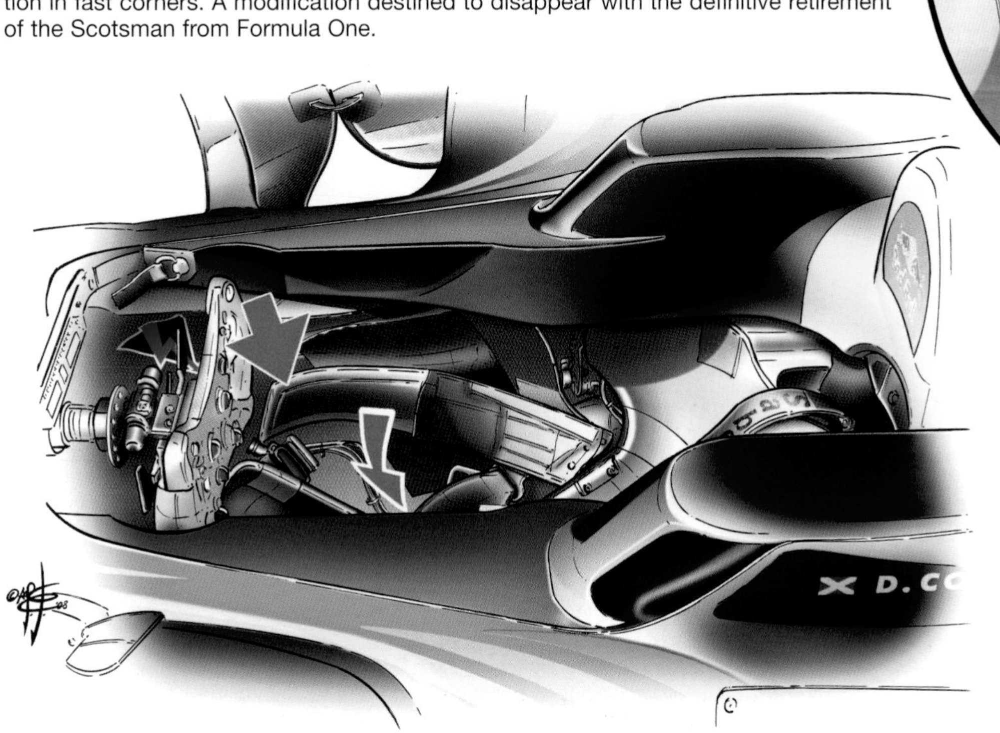

McLAREN

McLaren conserved the feature introduced to the display fitted into the dashboards in 2001 and not integrated into the steering wheel, as it was on most of the cars. Only Red Bull and Toro Rosso took the same path.

Talking about SUSPENSIONS

As in the 2007 season, aerodynamics played again a significant influence on all aspects of the cars, in particular on the suspension layouts, especially at the front end where the raising of the nose cones and the ever higher wishbone mounts obliged the designers to make notable compromises with respect to optimum kinematics. Effectively, the raising of the lower wishbone has led to a different strut angle, with significant modifications of the suspension loadings too.
Let's go back a short way. What has become the generalized suspension layout adopted on all cars involves double wishbones and reaction struts acting on a push-rod both at the front and the rear. In its various guises, the rocker itself acts on a link connected to an anti-roll bar, the purpose of which is to reduce roll movements and is equipped with a link to a third element with a spring and damper unit that controls vertical movements of the wheels (pitching), preventing the underbody from grounding during braking rather than over rough surfaces.
The rigidity of the suspension is controlled by torsion bars housed at the rocker fulcrum point. Varying their diameter varies the torsional rigidity and therefore the rigidity of the suspension as a whole.
This layout has been the subject of diverse interpretations in relation to the space available on the chassis, a factor itself determined above all by aerodynamic choices and restrictions.
There are therefore configurations

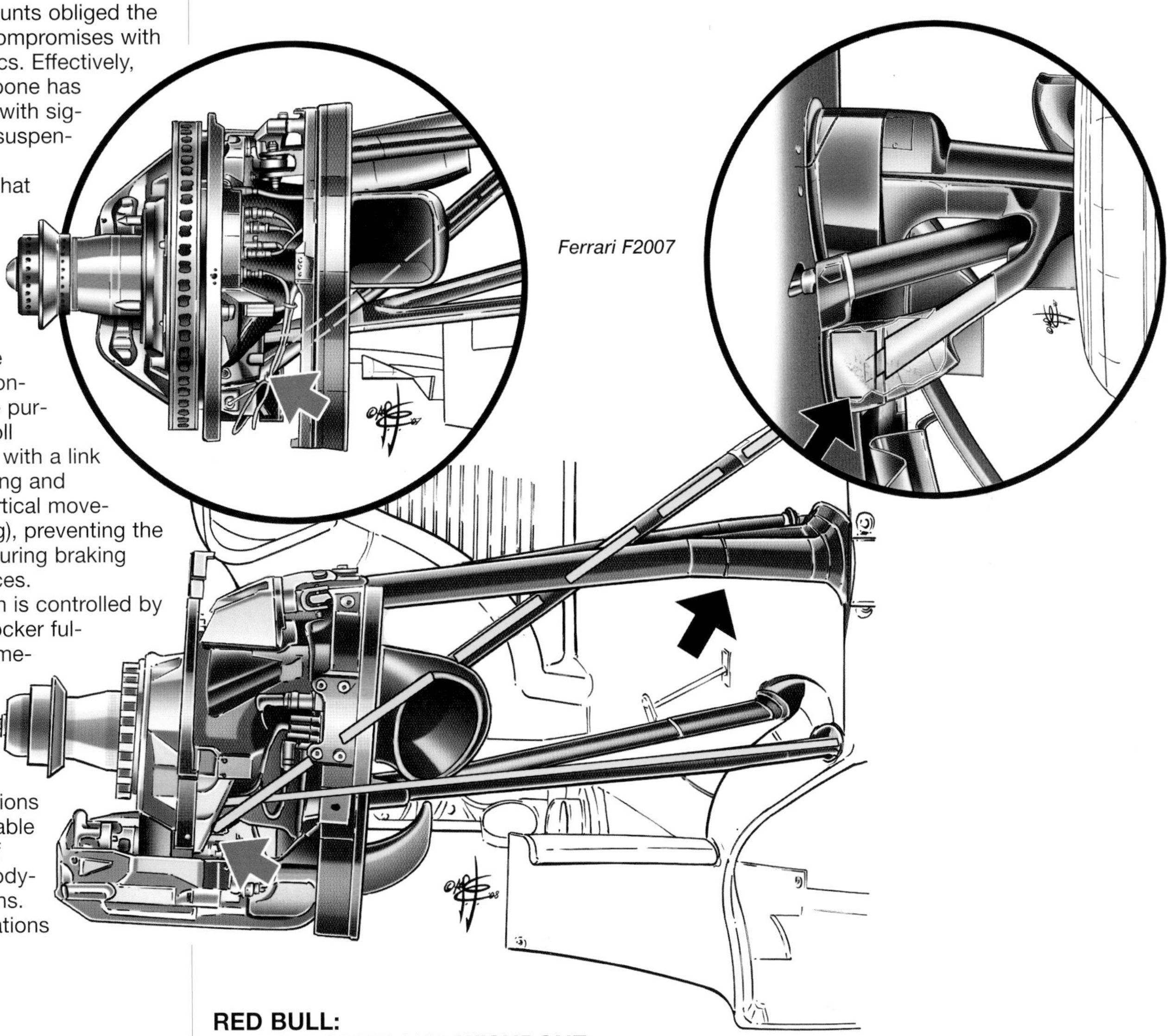

Ferrari F2007

RED BULL:
STRUT MOUNT AND WISHBONE
Ferrari introduced two further refinements in 2007: the anchoring of the push rod link not only directly on the wheel hub (a feature introduced in 2001) but at a point lower than the lower wishbone mount. It can clearly be seen how for aerodynamic motives the anchorage of the lower lever has been moved towards the centre of the hub and how the levers are anchored to the sides of the chassis and are therefore very short, the length compensating for the rigidity. Above all, we can see that the push rod link is anchored very low on the hub and that it actually passes through the lower wishbone and therefore maintains a very high angle with respect to the horizontal plane, tending to contain low forces, benefiting the structure and sections of the levers and therefore reducing unsprung weight.
Furthermore, with this configuration the lower lever is high enough for good aerodynamic efficiency.
Also of note was an element important in obtaining high suspension rigidity values. In order to increase the stiffness value of the levers and reduce their sections to improve aerodynamics, minimal-section elements were designed and manufactured in solid carbon fibre with unidirectional and woven high modulus M46J and M55J fibres.
In the circle on the right is another feature introduced by Ferrari and widely adopted on other cars: the anchoring of the upper wishbone to a structure on the outside of the chassis in order to reduce the length of the lever and achieve optimum camber recovery.

with horizontal rather than inclined bars mounted inside the front of the chassis or buried in the gearbox casting.
One variation on the theme has been seen in recent years with the introduction of the so-called Inerters or I-dampers, inertial dampers that are a direct evolution of the mass dampers considered to be illegal by the FIA and which replaced the third anti-dive element. McLaren was actually the first to use the feature during the era of the Renault scandal, the team having asked the Federation for clarification as it considered the Mass Damper to be an element external to the suspension system in contrast with the I-damper. This device improves suspension performance, in particular that of the dampers which, as is well known, are responsible for absorbing and damping the chassis movements provoked by irregularities in the road surface. Effective dampers allow contact to be maintained between tyre and road, reducing loss of grip and therefore improving directional control and traction out of corners.
When the driver rides the kerbs, the chassis movements provoked are extreme and cause the car literally to take off as the dampers are unable to absorb the impacts.
With the Inerter, part of the energy deriving from these impacts acts on the device linking the rockers through a threaded rod before it reaches the shock absorbers and is damped. Inside the device is a high-inertia flywheel that, at the moment of impact, thanks to the effect of the advancing linking rod, spins and disperses part of the force generated by the impact in the form of rotational kinetic energy. In this way the system, that functions like a spinning top, manages to increase the damping effect in parallel with the effect of the shock absorbers, guaranteeing optimum contact between tyre and asphalt and facilitating fluid driving through and out of chicanes.
The 2008 season saw this suspension feature adopted by every team, including Force India, with McLaren continuing to lead the way even with respect to its great rival Ferrari, in its second season with the I-damper.

RENAULT: MASS DAMPER

The I-damper reprises the inertial damping concepts of the mass damper banned by the Federation in 2006 after it had been at the centre of drawn-out arguments. In this case the application is considered to be legal because it is contained entirely within the layout of the suspension itself. The mass damper was, in fact, an external device, albeit connected to the suspension as can be seen in the illustration. It is clear that the mass (around 3.5 kg at the rear – indicated by the red arrow – and no less than 7 kg at the front, was physically outside the suspension assembly. The mass damper at the front controlled the variation in ride height, while the one at the rear, inside the gearbox, reduced the car's bouncing.

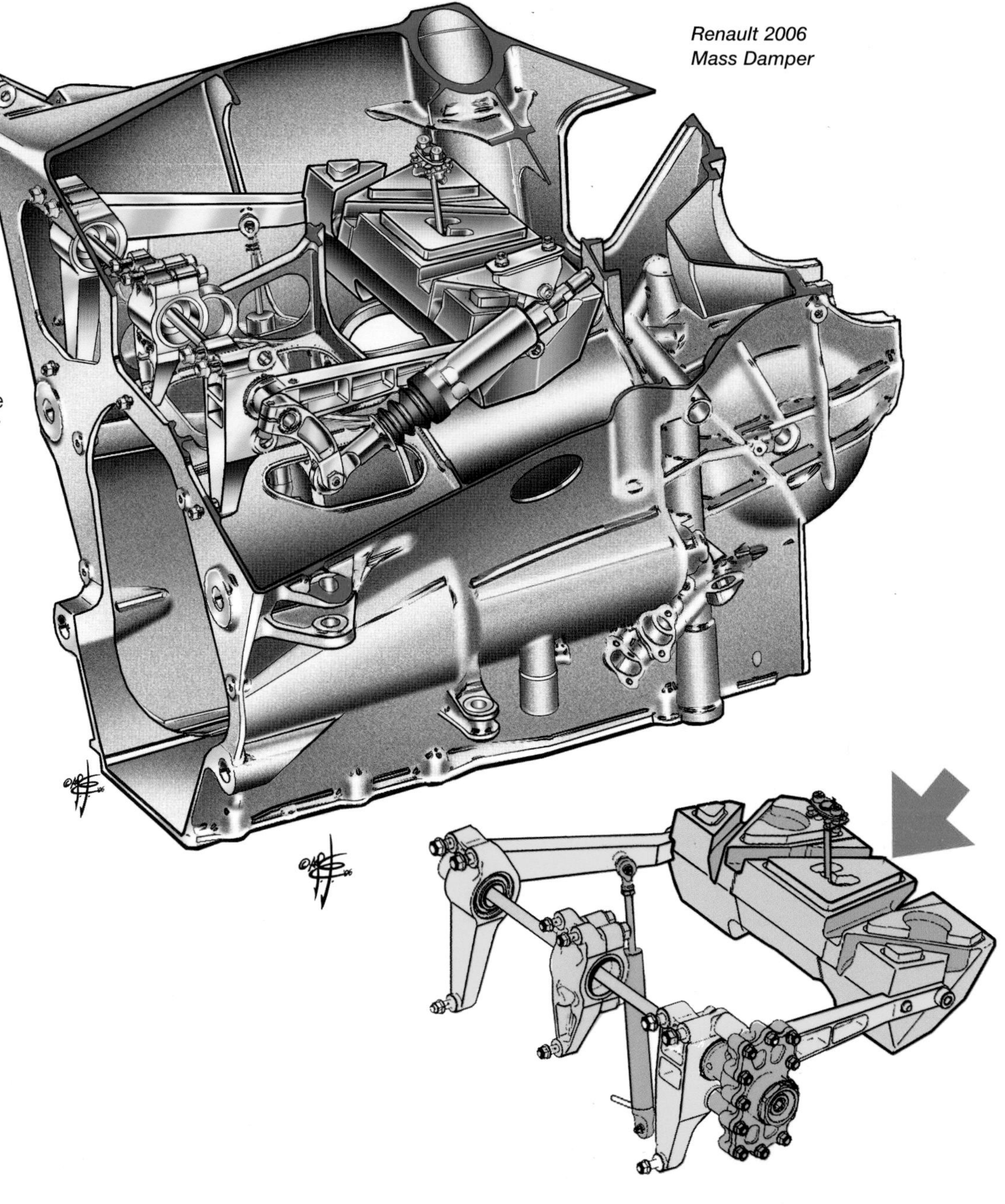

Renault 2006 Mass Damper

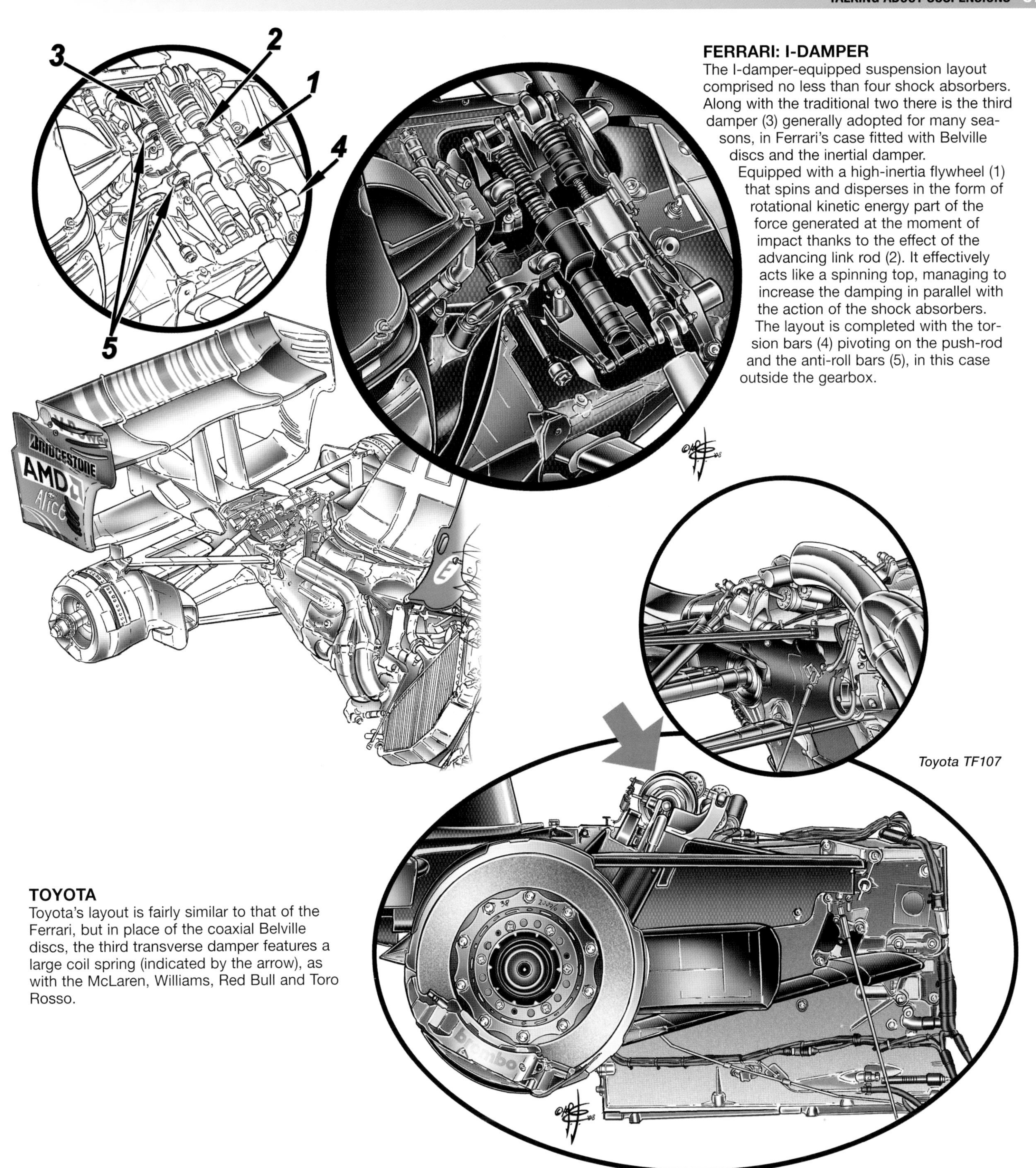

FERRARI: I-DAMPER

The I-damper-equipped suspension layout comprised no less than four shock absorbers. Along with the traditional two there is the third damper (3) generally adopted for many seasons, in Ferrari's case fitted with Belville discs and the inertial damper.

Equipped with a high-inertia flywheel (1) that spins and disperses in the form of rotational kinetic energy part of the force generated at the moment of impact thanks to the effect of the advancing link rod (2). It effectively acts like a spinning top, managing to increase the damping in parallel with the action of the shock absorbers.

The layout is completed with the torsion bars (4) pivoting on the push-rod and the anti-roll bars (5), in this case outside the gearbox.

Toyota TF107

TOYOTA

Toyota's layout is fairly similar to that of the Ferrari, but in place of the coaxial Belville discs, the third transverse damper features a large coil spring (indicated by the arrow), as with the McLaren, Williams, Red Bull and Toro Rosso.

The Ferrari F2008 was a great evolution of the F2007 (based on the version at the GP of Spain), its few defects revised and corrected.
First of all, the wheelbase was slightly shortened, while its rival McLaren's was lengthened, after which the Anglo-German car was still shorter than the F2008.
This allowed the move to the front the car's weight distribution without having to have considerable ballast in the nose, as had many other teams.
Aerodynamics were, once again, the principal inspiration of the entire project, attempting to produce a car with a constant yield to make the drivers' task easier as it was already difficult enough without traction control and engine brake.
At the car's launch, the most noticeable factor was the nose that was clearly inspired by Toyota, with that bridged flap elegantly linked to it in such a way as to create a net separation of air flows.
Starting from the GP of Spain, the nose was given that which became the most evident new feature of the season: the hole to create a passage of air from the lower and upper zones of the nose, about which more in the Aerodynamics chapter; the feature was used at all the high downforce circuits.
The original shape of the nose was rather interesting: it was higher from the ground compared with that of the F2007 and had a very concave and sophisticated lower section.
Despite the reduction in wheelbase, the sidepods were practically the same distance from the front axle in order to be able to neutralise with turning vanes the harmful turbulence that comes from the front wheels in movement. The intakes for cooling the radiators were raised and narrowed to create a more evident separation of the air flow in the lower part (jointly with the greatest break) that goes to feed the rear sector, which was also made more concave.
All of the rear end was miniaturised, with the exhausts slightly set back but closer to the car's body. The engine air intake was new, higher and narrower. In the

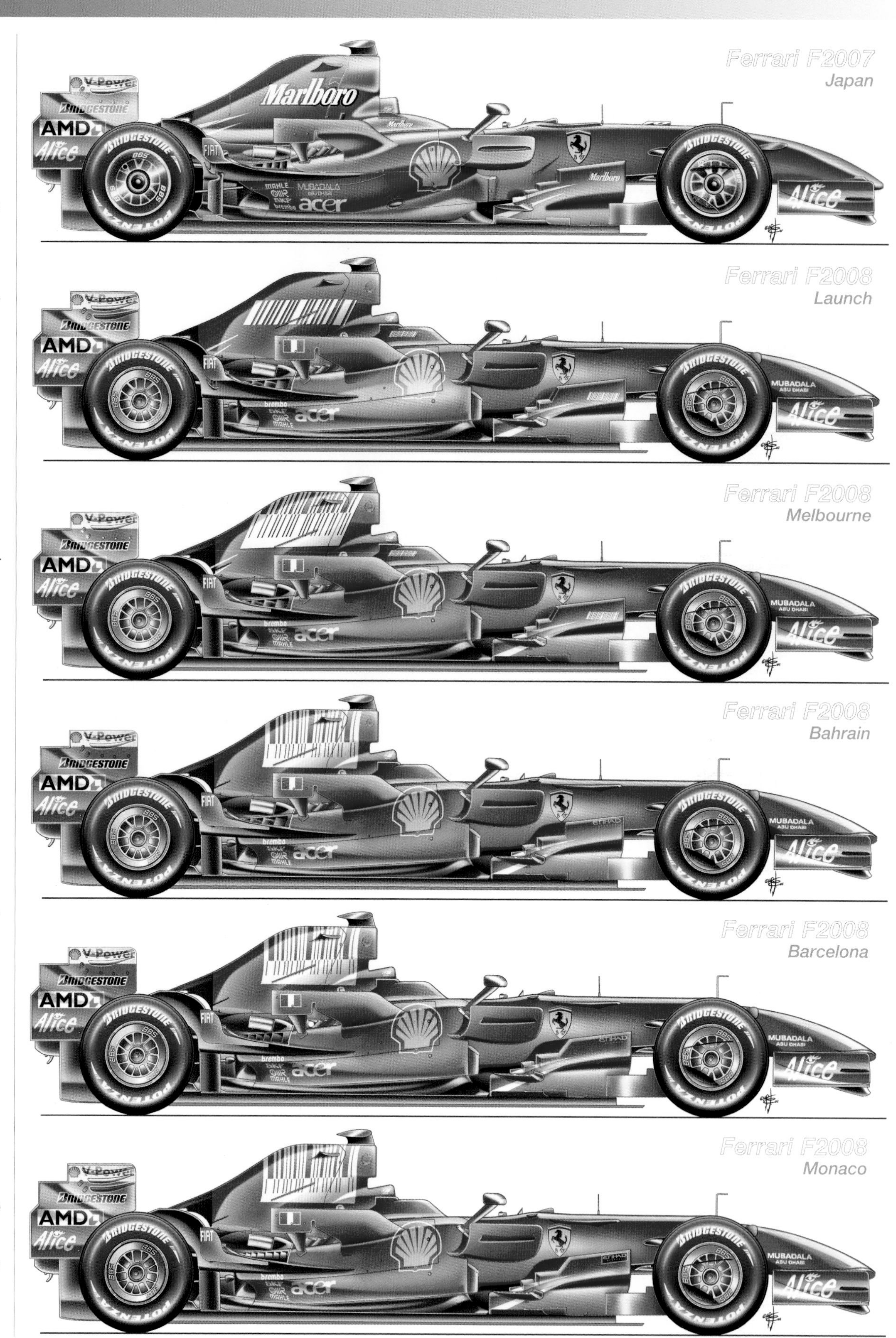

CONSTRUCTORS' CLASSIFICATION	2007	2008	
Position	1°	1°	= ■
Points	204	172	-32 ▼

construction of the chassis, Ferrari maintained the concept of the deformable structure fixed directly onto the stepped bottom, a concept that had a great influence on Williams.
The lower zone under the driver's legs was raised to ensure a greater flow of air to the diffuser.
On the suspension front, the team did a considerable amount of work on the layout that included the inertial dampers with no fewer than four units at the rear end (see the Suspensions chapter). During the season, the development of the car was incredible, with new elements at every race, the major part of which concerned the astonishing achievement in micro-development to adapt the car to the characteristics of the various circuits. While having introduced the new hole at the time of the car's design and not used by the other teams, Ferrari did adapt other cars' features to the F2008, like BMW's vertical link between the barge boards in the area of the start of the sidepods and the Red Bull cupola sail; two developments that became trends during the 2008 season.
The battle for the world championship between Ferrari and McLaren was, therefore, also a clash of micro-developments at every race.
There was the superiority in the exploitation of the suspension by McLaren, which was more at ease on the kerbs and with tyres in qualifying.
Ferrari replied with greater aerodynamic efficiency that gave it a kind of supremacy in fast corners. In racing, the Rossa permitted itself to select better performing tyres, being much "kinder" in their use over distance, while McLaren was often limited in this sector.
The F2008's tendency to understeer, with a weight distribution that was slightly further back than that of the MP4-23, penalised Raikkonen's driving style. In its defeat in the battle for the world drivers' title, the lesser reliability of the F2008 took its toll compared to its McLaren rival: four broken engines, Massa at Budapest and Raikkonen at Valencia, plus one of the Finn's exhausts broken at Magny Cours as well as a series of unfavourable circumstances that even culminated in mistakes in the pits. Despite all of that, in the end the fight for the constructors' title saw a margin of 21 points to Ferrari's advantage.

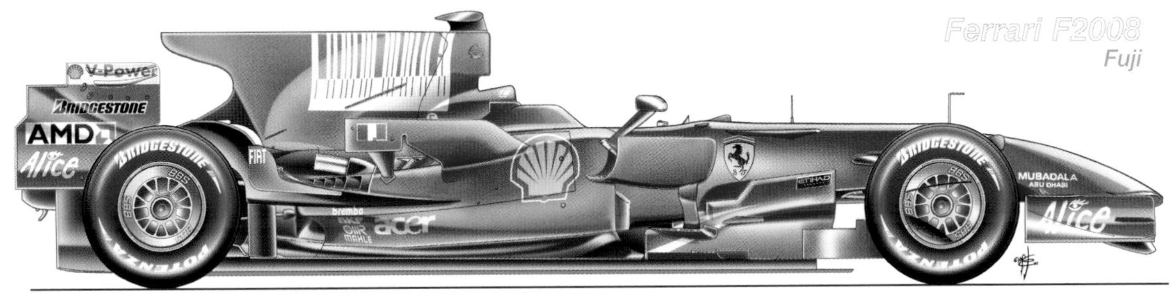

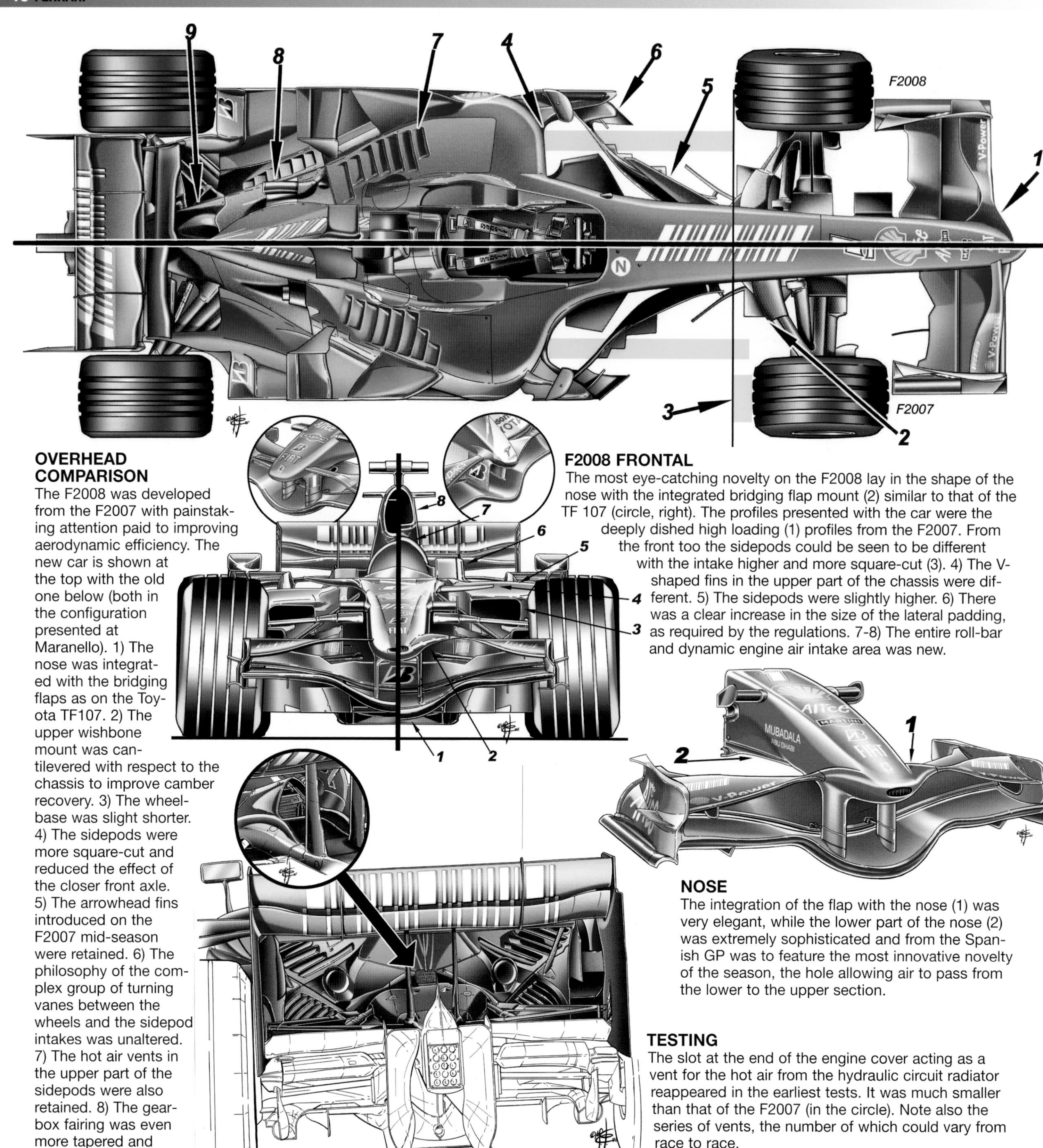

OVERHEAD COMPARISON

The F2008 was developed from the F2007 with painstaking attention paid to improving aerodynamic efficiency. The new car is shown at the top with the old one below (both in the configuration presented at Maranello). 1) The nose was integrated with the bridging flaps as on the Toyota TF107. 2) The upper wishbone mount was cantilevered with respect to the chassis to improve camber recovery. 3) The wheelbase was slight shorter. 4) The sidepods were more square-cut and reduced the effect of the closer front axle. 5) The arrowhead fins introduced on the F2007 mid-season were retained. 6) The philosophy of the complex group of turning vanes between the wheels and the sidepod intakes was unaltered. 7) The hot air vents in the upper part of the sidepods were also retained. 8) The gearbox fairing was even more tapered and closely fitting.

F2008 FRONTAL

The most eye-catching novelty on the F2008 lay in the shape of the nose with the integrated bridging flap mount (2) similar to that of the TF 107 (circle, right). The profiles presented with the car were the deeply dished high loading (1) profiles from the F2007. From the front too the sidepods could be seen to be different with the intake higher and more square-cut (3). 4) The V-shaped fins in the upper part of the chassis were different. 5) The sidepods were slightly higher. 6) There was a clear increase in the size of the lateral padding, as required by the regulations. 7-8) The entire roll-bar and dynamic engine air intake area was new.

NOSE

The integration of the flap with the nose (1) was very elegant, while the lower part of the nose (2) was extremely sophisticated and from the Spanish GP was to feature the most innovative novelty of the season, the hole allowing air to pass from the lower to the upper section.

TESTING

The slot at the end of the engine cover acting as a vent for the hot air from the hydraulic circuit radiator reappeared in the earliest tests. It was much smaller than that of the F2007 (in the circle). Note also the series of vents, the number of which could vary from race to race.

NOSE
A comparison between the F2008 nose and an early depiction of the hole for the passage of air from the lower to the upper section. News of this feature had in fact filtered through even before the championship got underway. Compared with reality, here the slot was set further forwards without altering the shape of the chassis, as instead was the case with the actual feature introduced in Barcelona.

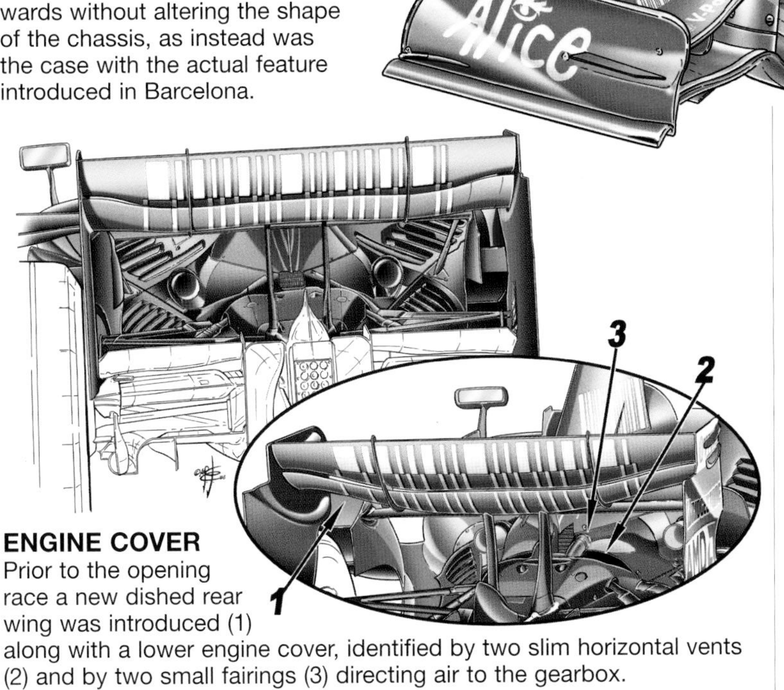

ENGINE COVER
Prior to the opening race a new dished rear wing was introduced (1) along with a lower engine cover, identified by two slim horizontal vents (2) and by two small fairings (3) directing air to the gearbox.

TURNING VANES
The small turning vanes set within the suspension were integrated with their mount to form a single arching element partially inspired by that of the 2007 McLaren.

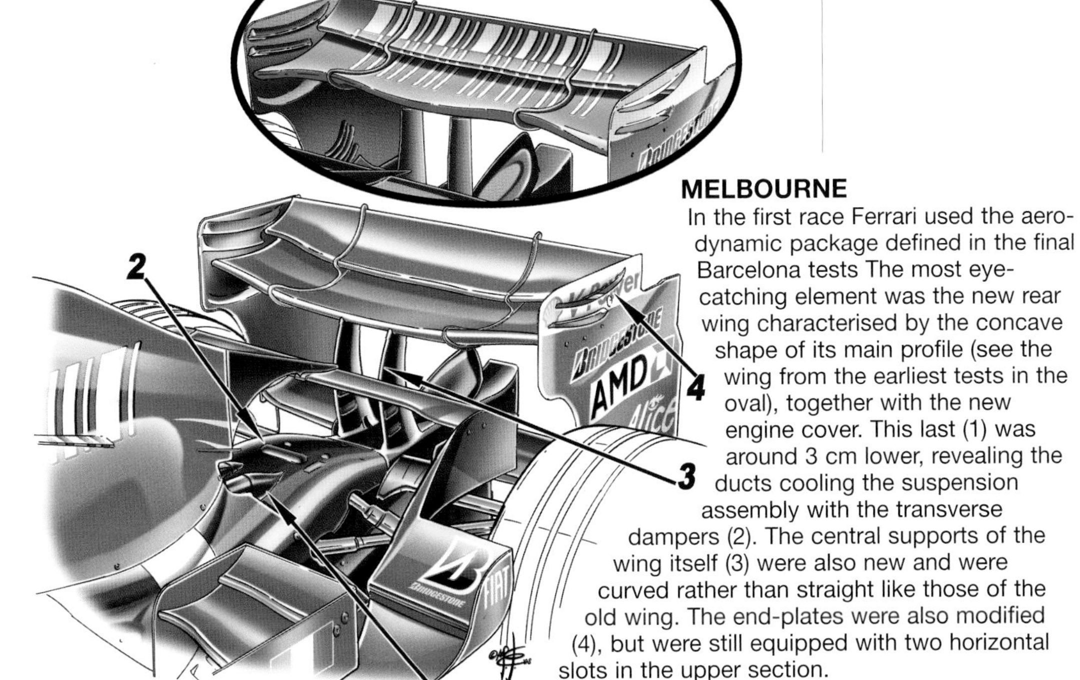

MELBOURNE
In the first race Ferrari used the aerodynamic package defined in the final Barcelona tests The most eye-catching element was the new rear wing characterised by the concave shape of its main profile (see the wing from the earliest tests in the oval), together with the new engine cover. This last (1) was around 3 cm lower, revealing the ducts cooling the suspension assembly with the transverse dampers (2). The central supports of the wing itself (3) were also new and were curved rather than straight like those of the old wing. The end-plates were also modified (4), but were still equipped with two horizontal slots in the upper section.

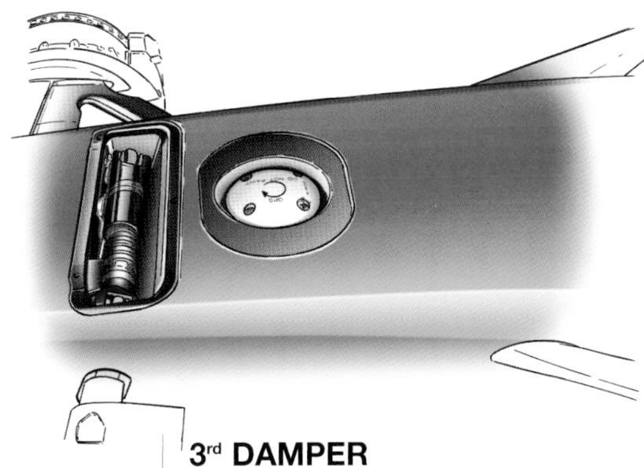

3rd DAMPER
Seen here is the third damper equipped with Belleville discs controlling the ride height and roll in the niche immediately behind the GPS used to signal the car's position on the track.

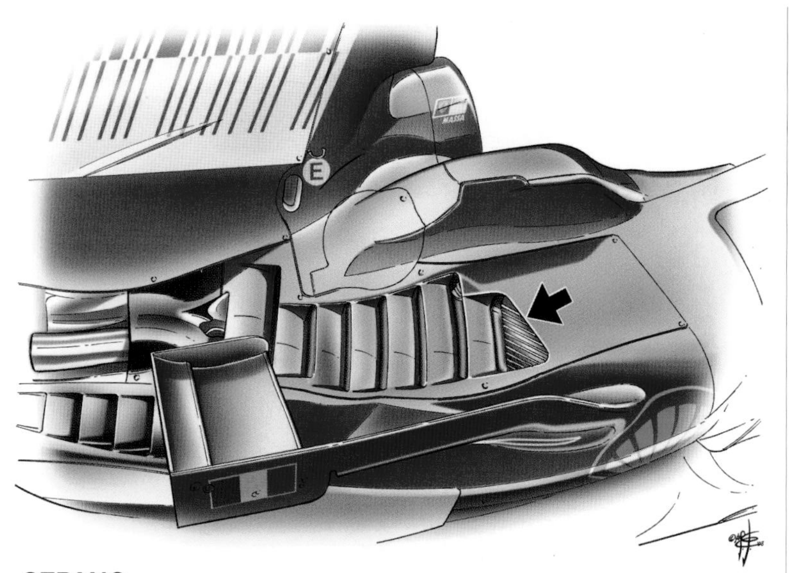

SEPANG
Ferrari had no need to use the version with the enlarged first cooling slot as it had in Melbourne given that in Malaysia the ambient temperature proved to be unusually lower than predicted; it therefore raced with the same number of slots but in standard form.

SAKHIR
A new rear wing flap set with a lower angle of incidence and slightly higher, as can be seen in this detail. Massa used this variant for qualifying and the race, while Raikkonen preferred the combination of the two profiles in order to obtain greater loading, a choice that proved to be less effective.

DEFORMABLE STRUCTURES
Ferrari had already modified the system of mounting the lateral deformable structures the previous season. They were no longer fixed to the chassis but rather attached to the stepped bottom that was then used in the crash test. In this way the same stiffness is achieved with a certain saving in weight.

FERRARI
The two Ferrari drivers used slightly different aerodynamic configurations. On the Friday only Massa had the new mini-flaps on the chimney without the vent, while Raikkonen tried those already seen in the first two races. Subsequently the Finn went for the new version but set his rear wing for a higher load.

FERRARI
A new flap on the rear wing offering less downforce combined with the new main profile already seen in the first two races. Note the new livery, with a significant white element on the main profile.

BARCELONA

Ferrari's new nose, hinted at early in the season, debuted with a one-two victory. In effect, the hole in the lower section had a shape mid-way between a NACA intake and an oval and represented an extension of the central section of the flap. It acted as an extractor in this high pressure area, avoiding the detachment of the air flow and therefore improving the efficiency of the front wing. The channelled air is then expelled from the upper section with a small flap inserted in the exit vent. The air flow does not disturb the driver as it effectively passes over his head. It can clearly be seen how the chassis was cut away during the design phase in order to accommodate the second slot, with the furrow covered in the early races by a kind of cover that prevented us from locating the exact position of the vent.

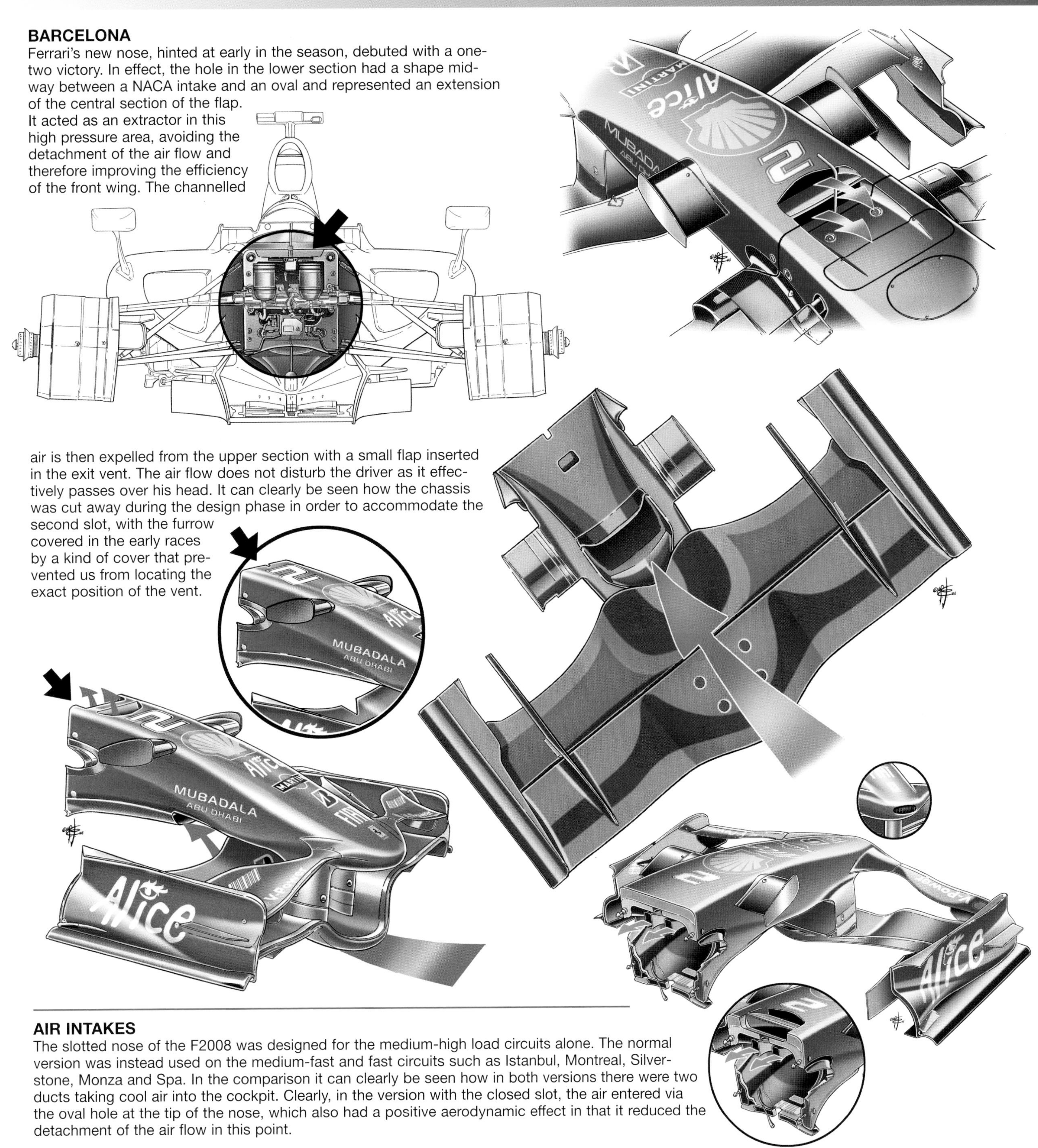

AIR INTAKES

The slotted nose of the F2008 was designed for the medium-high load circuits alone. The normal version was instead used on the medium-fast and fast circuits such as Istanbul, Montreal, Silverstone, Monza and Spa. In the comparison it can clearly be seen how in both versions there were two ducts taking cool air into the cockpit. Clearly, in the version with the closed slot, the air entered via the oval hole at the tip of the nose, which also had a positive aerodynamic effect in that it reduced the detachment of the air flow in this point.

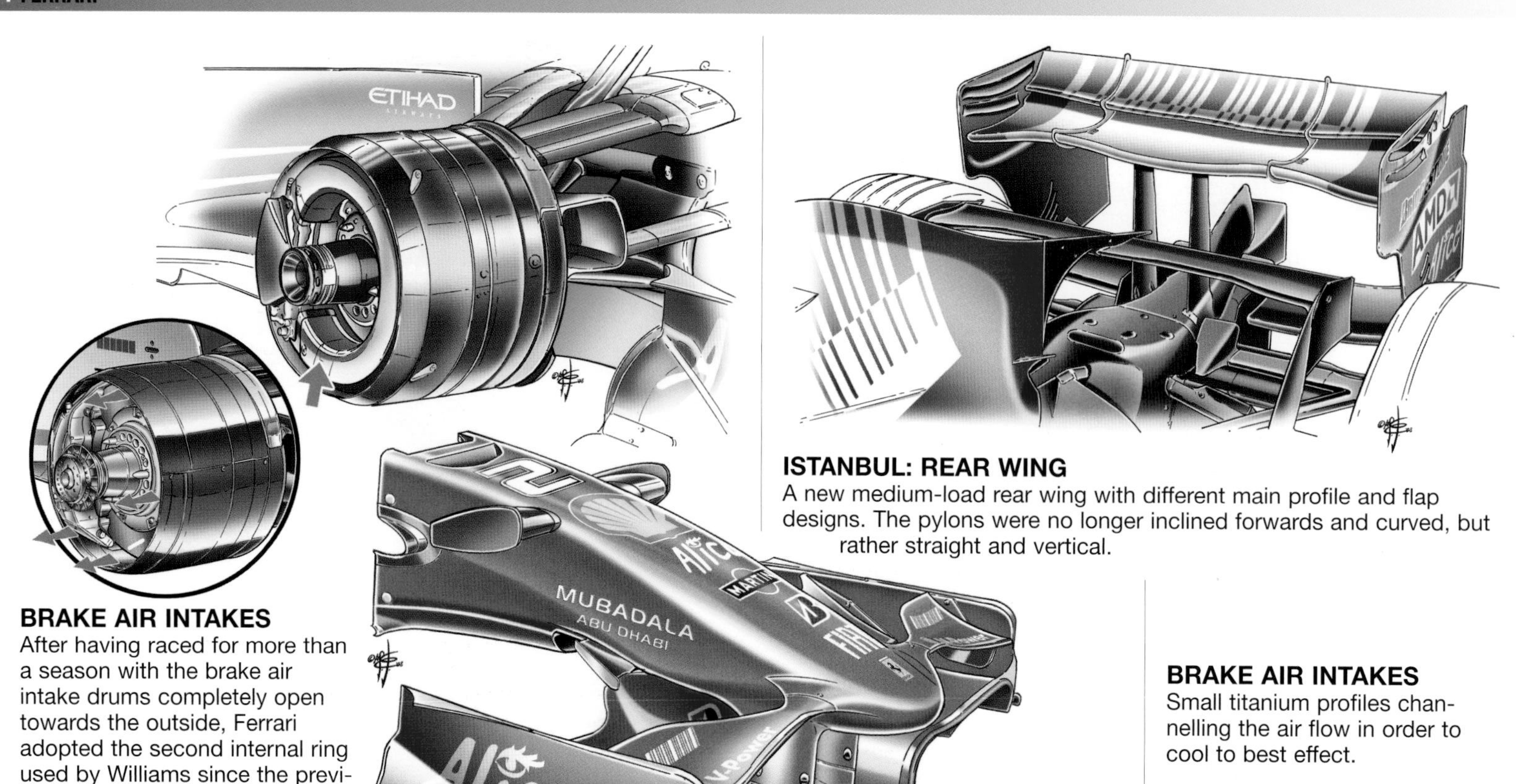

ISTANBUL: REAR WING
A new medium-load rear wing with different main profile and flap designs. The pylons were no longer inclined forwards and curved, but rather straight and vertical.

BRAKE AIR INTAKES
After having raced for more than a season with the brake air intake drums completely open towards the outside, Ferrari adopted the second internal ring used by Williams since the previous season and adopted by BMW in '08, leaving a narrow space between the two elements to expel the hot air outwards. It should also be noted that the front wheel covers set a trend in Spain, with Red Bull and Renault following Ferrari and McLaren's lead, while for the moment BMW, Honda and Toyota forewent the feature.

MONACO
On its second outing the slotted nose of the F 2008 was modified with the addition of two small spoilers either side of the intake mouth in a continuation of the new flap. The end-plates were slightly cropped to prevent the greater steering angle (around 22°) from damaging the tyres.

BRAKE AIR INTAKES
Small titanium profiles channelling the air flow in order to cool to best effect.

MONACO: REAR WING
In contrast with the McLaren policy, Ferrari exploited the full potential volume for the rear wing profiles with the main plane being straighter and combined with a flap with a notable angle of incidence and a deeper chord in the central section. The end-plates had three slots in place of the two of the Turkish GP version.

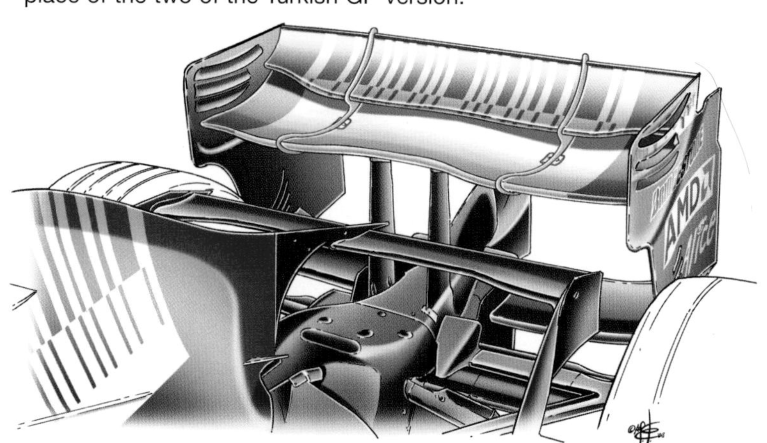

VENTS
The low speeds of Monaco made heat dispersal difficult, Ferrari's response being to increase the number of vents on the sidepods, with a series also being let symmetrically into the part inboard of the chimneys.

MONTREAL

A new front wing characterised by a flap (with a cropped central section) providing a lower aerodynamic load and that did not interfere with the braking system's greater cooling requirements.

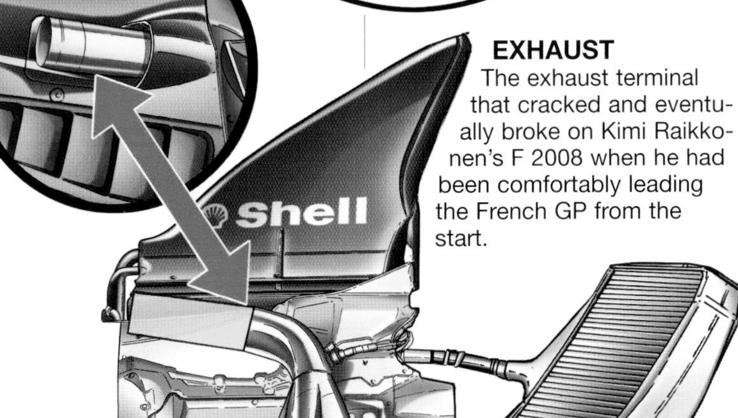

EXHAUST

The exhaust terminal that cracked and eventually broke on Kimi Raikkonen's F 2008 when he had been comfortably leading the French GP from the start.

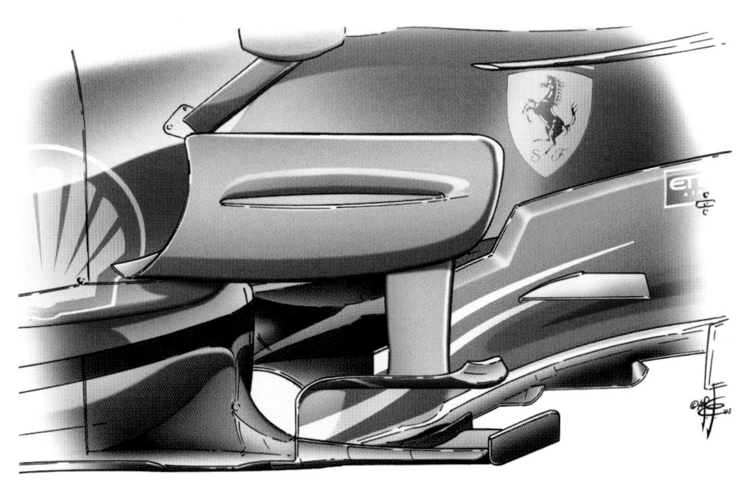

HOCKENHEIM

At Hockenheim a further 2 teams (Ferrari and Honda) adopted the vertical link (introduced by BMW from the start of the season) between the lateral boomerangs and the large lower bargeboards to better channel the flow of air towards the underside of the car, cleaning up the turbulence generated by the front wheels. Up to 15% of the total downforce can be created in this area. Only 2 teams resisted following this trend: McLaren and Force India.

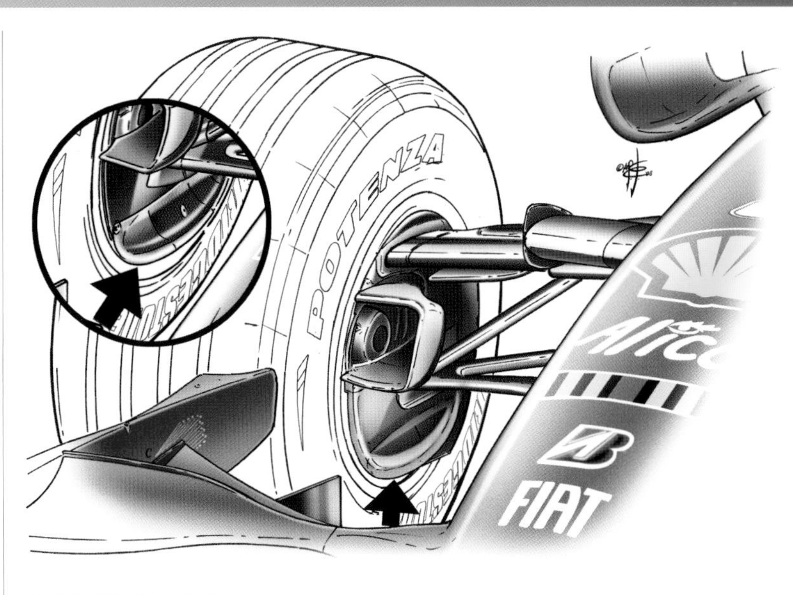

MAGNY COURS

New brake air intake on the F 2008. The modification concerned the banana-shaped internal section that was less accentuated with respect to the configuration used from the German GP in 2007 and was designed to optimise the flow in the channel between wheel and chassis.

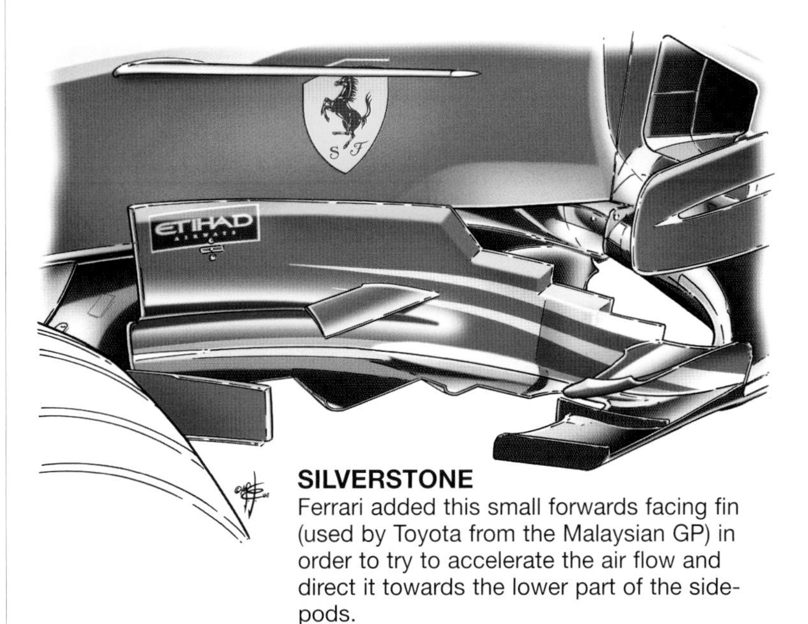

SILVERSTONE

Ferrari added this small forwards facing fin (used by Toyota from the Malaysian GP) in order to try to accelerate the air flow and direct it towards the lower part of the sidepods.

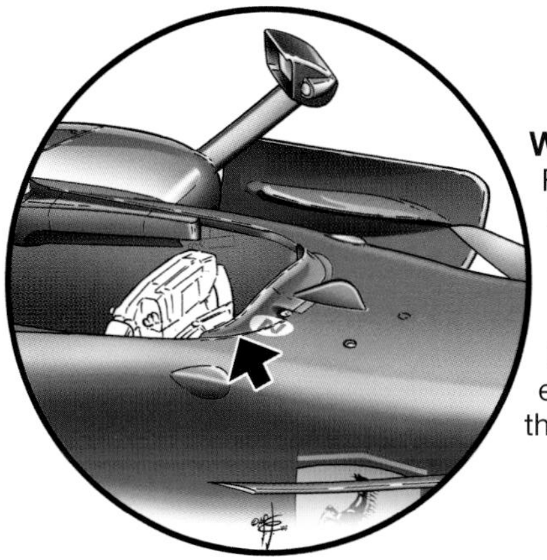

WINDSCREEN

Ferrari modified the windscreen area, initially vertical and set around the chassis aperture. It was now replaced with a small upturned lip that reduced the length of the aperture and limited the turbulence in the area of the driver's head.

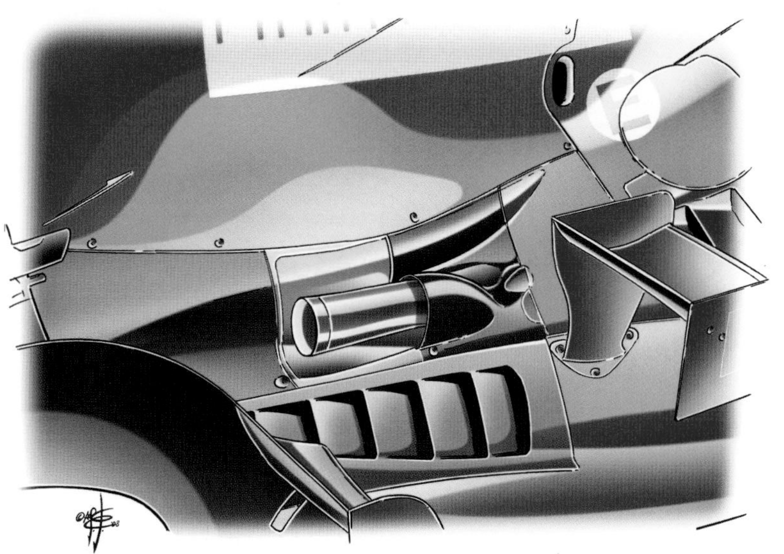

ENGINE COVER
New bodywork in the exhaust area, with greater evacuation of the hot air from inside the sidepods with respect to the version used up to this point. The exhaust terminal remained the same length.

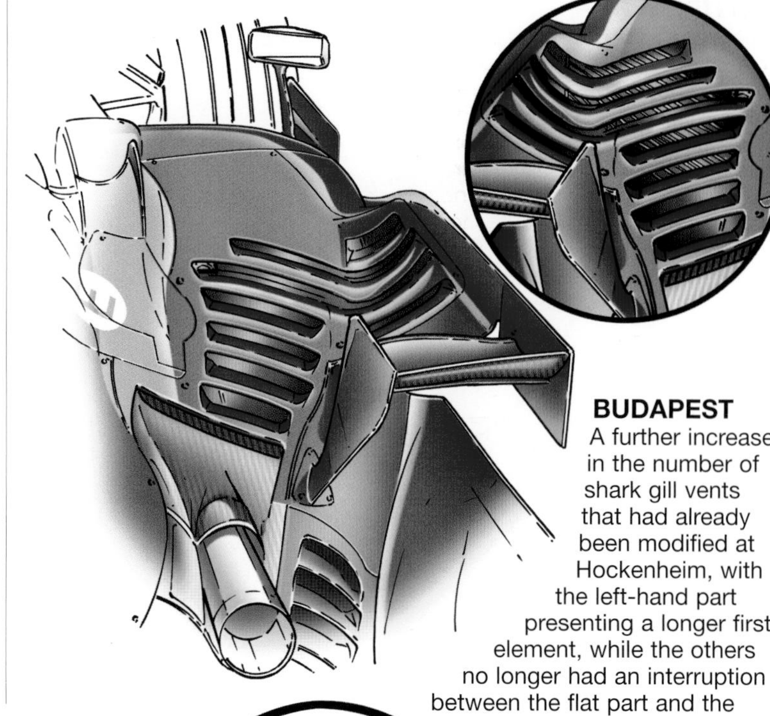

BUDAPEST
A further increase in the number of shark gill vents that had already been modified at Hockenheim, with the left-hand part presenting a longer first element, while the others no longer had an interruption between the flat part and the part inboard of the chimneys.

LARGE FIN
The debut of the large dorsal fin that helped balance the car better with greater rear wing efficiency and reduced drag with respect to the same aerodynamic load.

BRAKE AIR INTAKES
In order to improve heat dispersal, the slot (introduced in Monaco) in the external area between the two drums was enlarged, while the intake mouth inside the wheels was left unchanged.

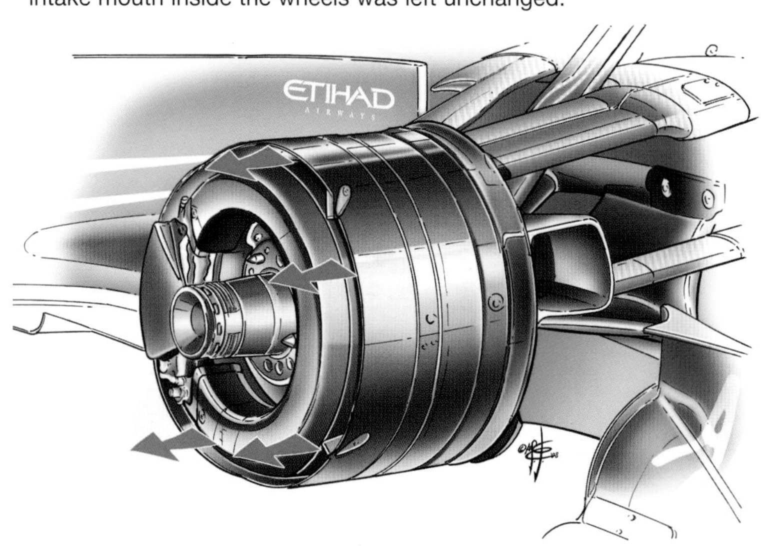

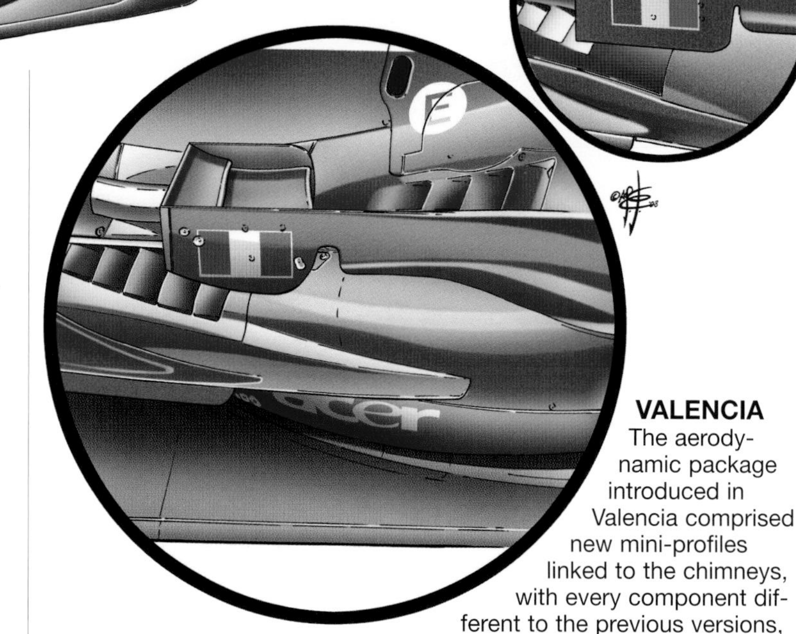

VALENCIA
The aerodynamic package introduced in Valencia comprised new mini-profiles linked to the chimneys, with every component different to the previous versions, from the profiles (less downforce) to the end-plates, the more rounded shape of which immediately distinguished it from its predecessor.

"LETTER BOX"

After debuting in Japan around a year earlier (although they had already been present at Spa on the unused spare car) the so-called "letter-box" end-plates were eliminated. They had been characterised by a long horizontal slot (1). The more curvilinear vertical link (2) was also new and the fin (3) was different, shorter and trapezoidal in form. The role of this element was to ensure that the air flow remained attached at that point of the bargeboards.

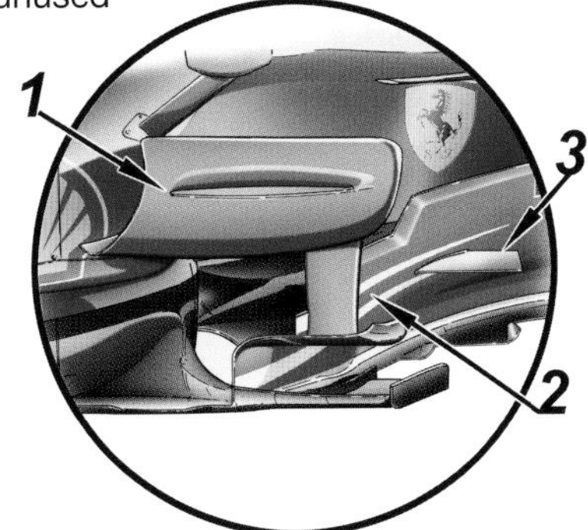

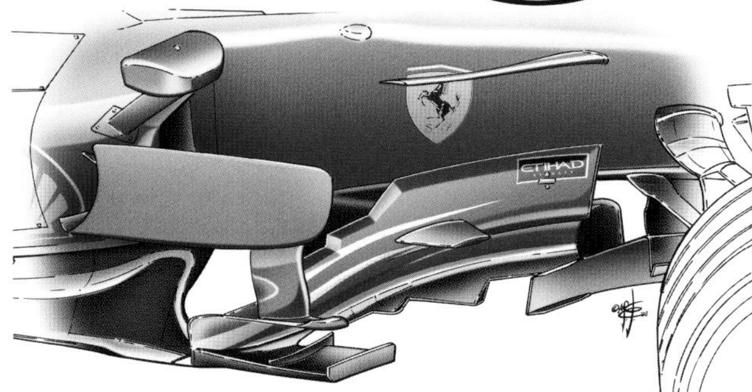

SPA

After having tested two new front wings on the Friday, Ferrari opted to race at Spa with the version with traditional vertical pylons, introducing however the small fin stabilizing the air flow towards the central section of the car. The version with the new cut-away supports guaranteeing a more efficient main profile was due to be used at Monza.

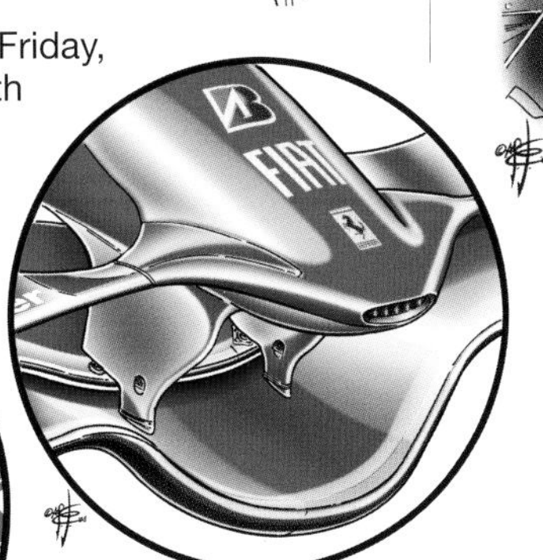

FERRARI

At the pit stop Ferrari removed the front wheel covers, firstly on Raikkonen's car and then on Massa's in order to improve brake disc cooling, thus leaving exposed the long coupling hub required for this feature (introduced at Silverstone in 2007).

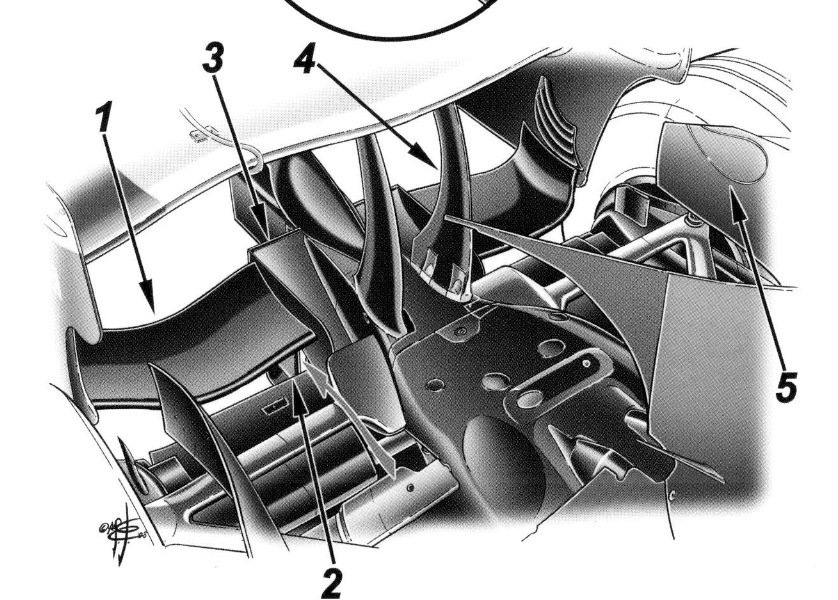

REAR END

The Ferrari rear end as seen at Spa was a compendium of new features. The most eye-catching of these was the particularly sinuous profile at the height of the gearbox (1) (very similar to that of the McLaren). The central extractor channel had a vertical vent (2) to avoid stalling, while the chute section was also modified (3). The vertical supports were new (4) with a reduced chord in the area in the vicinity of the main profile mount. The reduced-chord, metre-wide profile (5) attached to the fins inside the rear wheels was removed.

MONZA: REAR WING

The F2008s reappeared with the original asymmetric configuration introduced the previous year, with the main profile having a longer chord (1) on the right to compensate for the shadow effect generated by the bargeboards (2) in fast right-hand corners. The new profile (3) introduced in Belgium at the height of the deformable structure was retained, while the wide profile (4) at the height of the rear axle was eliminated. The chord of the main profile was, however, increased with respect to the flap to guarantee greater stability under braking.

SINGAPORE

Only Massa used the new shaped supports, this time applied to the slotted nose. Both drivers instead used an evolution of the fin applied either side of the intake mouth of the slot introduced in Monaco: in Singapore it had a delta shape.

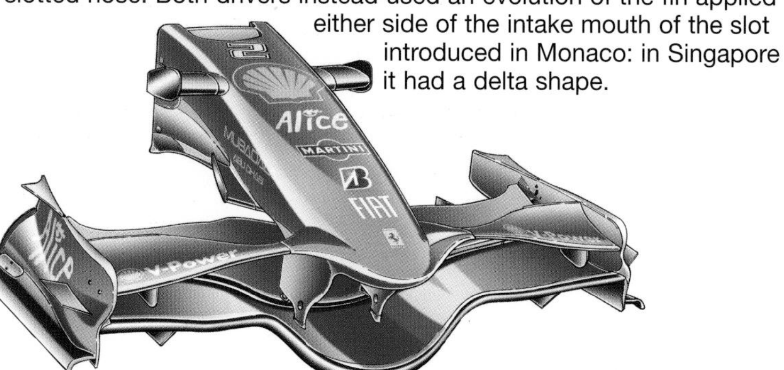

BRAKE AIR INTAKES

New cooling air intakes with an enhanced aerodynamic function in cleaning the air flow in this area, as can be seen from the greater surface area of the vertical fin at the rear.

SAO PAOLO

After having completed the previous 3 races with different front wing supports, the Ferrari drivers made the same choice in Brazil. Massa in fact went for the wing used previously by Raikkonen, with the old vertical rather than curved supports. The new configuration (introduced in Belgium and raced by Massa from the Singapore GP) guaranteed greater loading but was more critical. The Ferrari team as a whole thus preferred the old configuration in Brazil as it provided greater consistency in the balancing of the car, an important factor on such an undulating track as Sao Paolo.

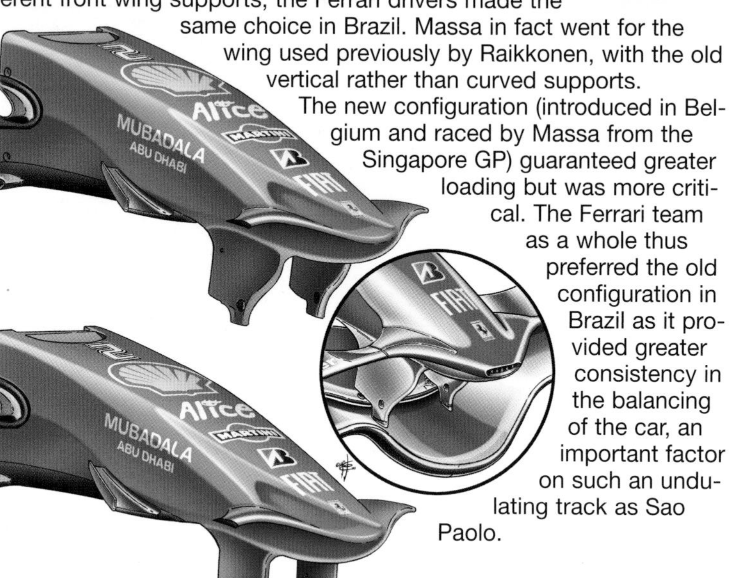

FUJI

There were different configurations for the two Ferraris, with Massa opting, as in the previous two races, for the new front wing support, while Raikkonen retained the old one, but also different engine covers for qualifying and the race. Raikkonen preferred the version without fins, the absence of which allowed mid-corner understeer to be reduced, while Massa opted for the fin that allows for increased rear wing downforce capacity when the car is yawing.

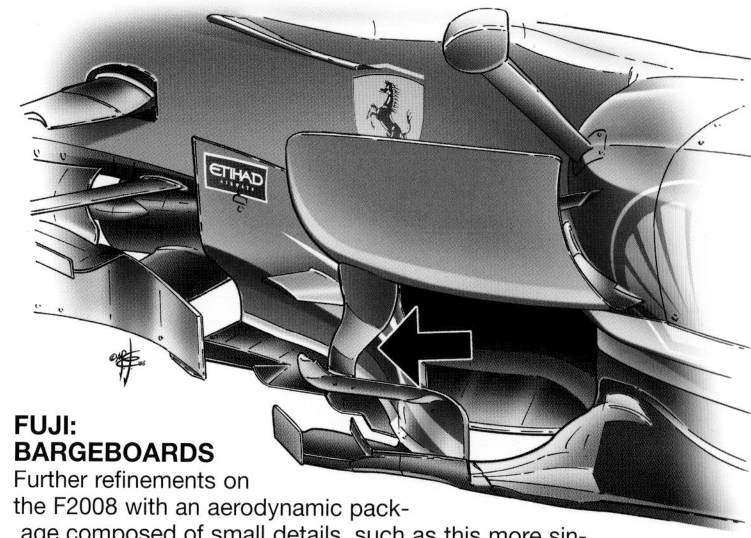

FUJI: BARGEBOARDS

Further refinements on the F2008 with an aerodynamic package composed of small details, such as this more sinuous and twisting vertical link between the boomerangs and the turning vanes designed to improve the quality of the air flow towards the rear section of the car.

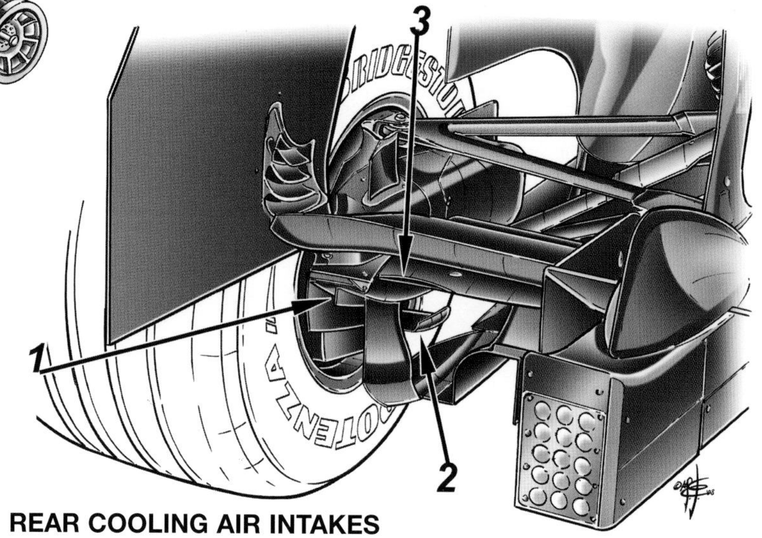

REAR COOLING AIR INTAKES

There were also new rear brake cooling intakes for the Ferrari in Singapore, with the configuration being retained for the following two races. Along with the fin indicated by the number (2), introduced mid-season, a small vertical fin appeared (1). Note the small Gurney flap (3) applied to the tie-rod fairing. These aerodynamic appendices applied to the brake cooling air intakes are currently being discussed in relation to the 2009 regulations and could be one of the few areas of significant development for the teams' new cars, stripped of those aerodynamic devices that saw their maximum development in the 2008 season.

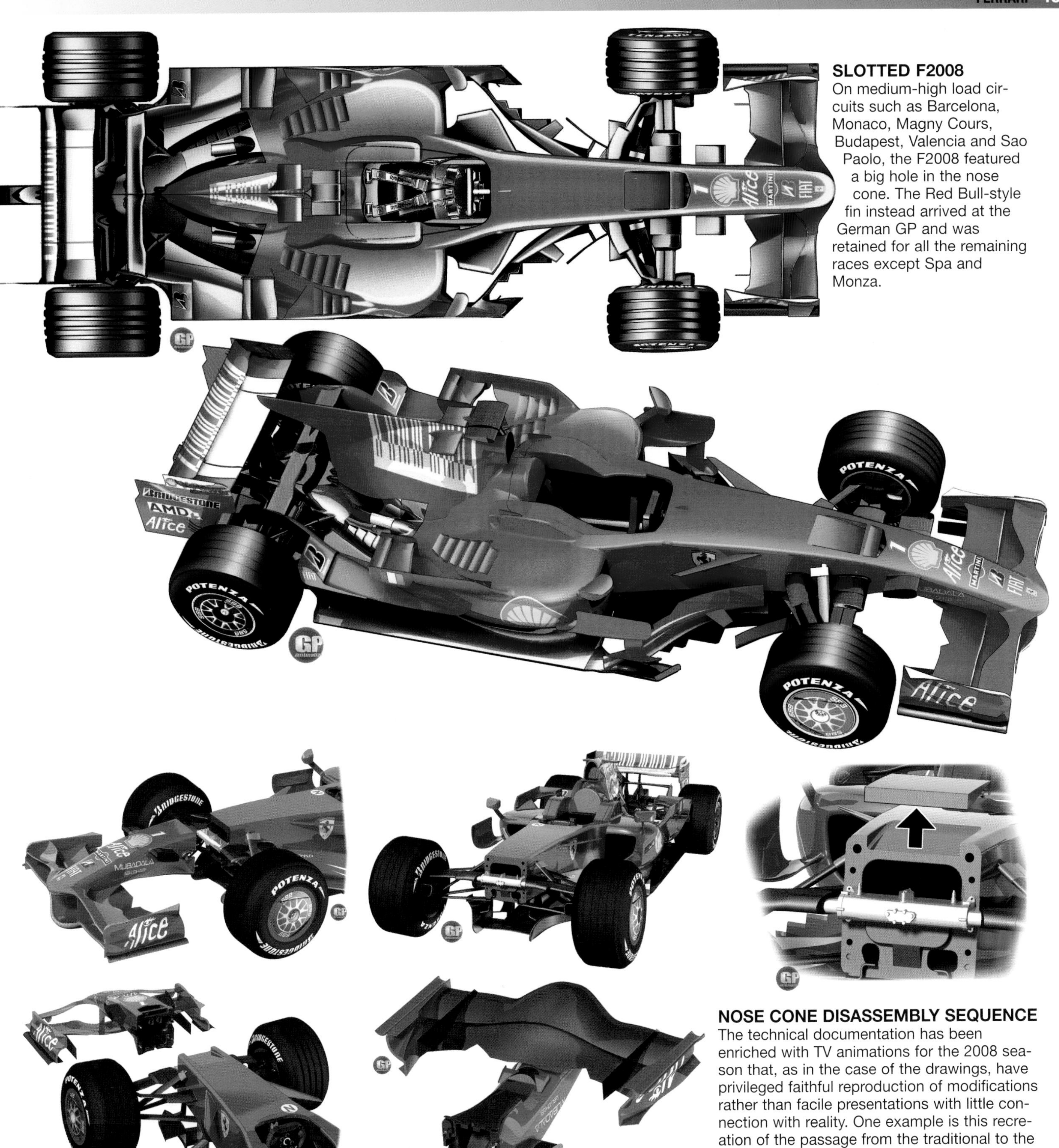

SLOTTED F2008

On medium-high load circuits such as Barcelona, Monaco, Magny Cours, Budapest, Valencia and Sao Paolo, the F2008 featured a big hole in the nose cone. The Red Bull-style fin instead arrived at the German GP and was retained for all the remaining races except Spa and Monza.

NOSE CONE DISASSEMBLY SEQUENCE

The technical documentation has been enriched with TV animations for the 2008 season that, as in the case of the drawings, have privileged faithful reproduction of modifications rather than facile presentations with little connection with reality. One example is this recreation of the passage from the traditional to the slotted nose produced in collaboration with Generoso Annunziata.

F2007-F2008 COMPARISON

The profile comparison between the two stripped cars clearly shows how the F2008 is a revised and updated version of the F2007 of which it retained many aspects. The most important difference centres on a slight reduction in the wheelbase that is almost imperceptible at the scale of the drawings. Great attention was paid to the refinement of the aerodynamics in a search for optimum efficiency, an area in which the F2008 bettered its MP4-23 rival.

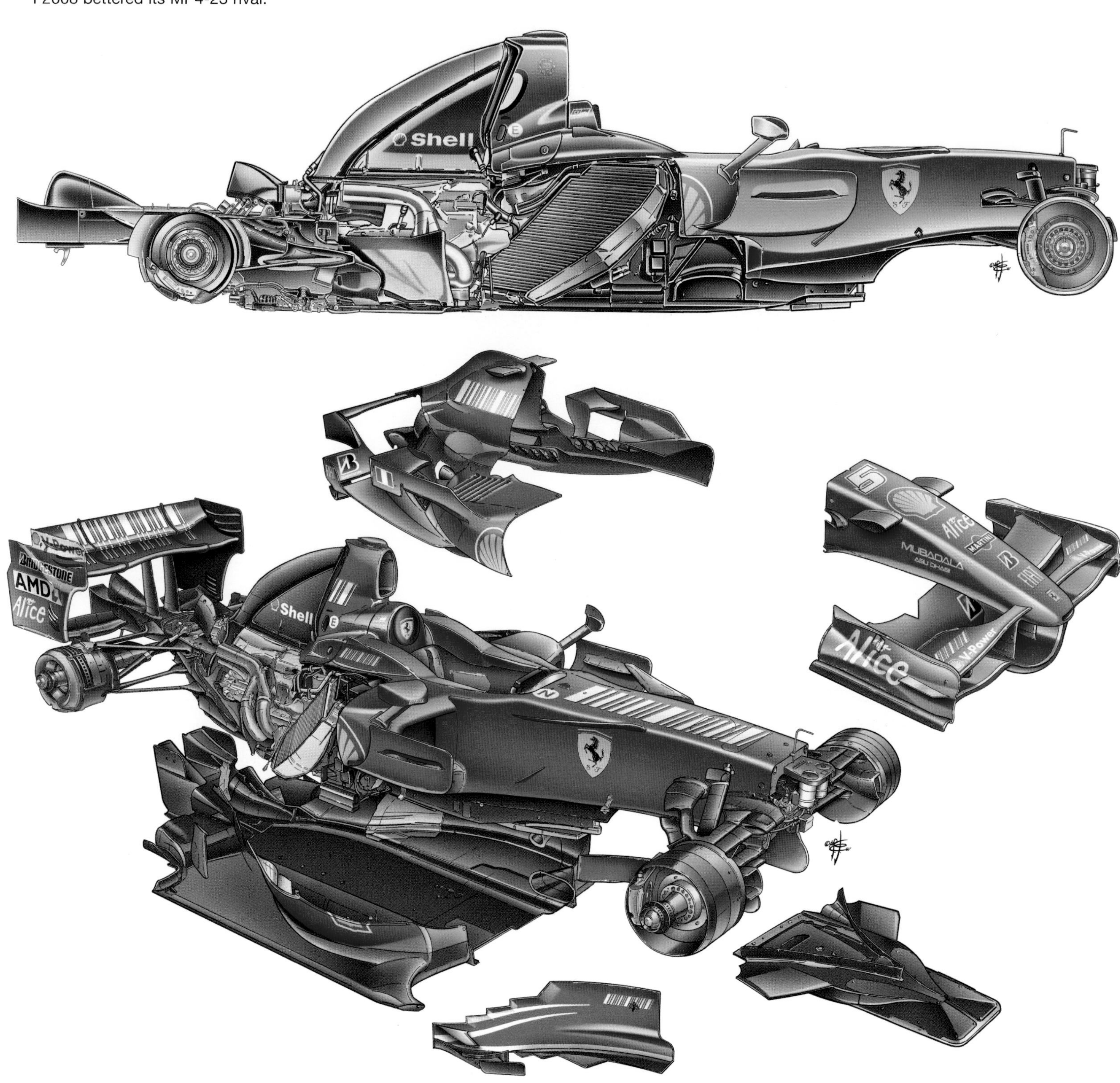

Shell
Bridgestone
AMD
Alice
Shell
MUBADALA
FIAT
V-Power
Alice
ETIHAD

The MP4-23 was the logical evolution of the car that battled with Ferrari for the world drivers' title in 2007. At the launch, the front wing was still that of the 'old' MP4-22 – as was the case with many others. The familiar area started immediately at the nose, which was straight and inclined downward but less like a duck's neck, and the planes stayed at three to exploit all the dimensions of the chord permitted by the regulations, even if one must say that in that sector McLaren fielded something completely new, the four plane became five (with a central blower) on high downforce tracks. The bridged flap was also given a slot to produce a more aggressive effect; it was the McLaren answer to the major new development of the season: the new hole in the Ferrari nose. The MP4-23 had a slightly longer wheelbase to push to the full the exploitation of its aerodynamics, following the work carried out by Ferrari in the 2007 season which, curiously, had slightly shortened that of the F2008. Another new feature was in the roll bar area, where with great structural difficulty McLaren were able to position a vertical air vent, which took the place of the previous season's Viking horns, which were actually dropped from the 2007 GP of Germany.
The solution of the chimneys integrated with the barge boards placed at the start of the sidepods was elegant and efficient and had the task of better directing the air flow toward the rear of the car. The MP4-23's chimneys also carried out, however, the task of dissipating the heat of the sidepods. The rear end retained the elegant and tapered shape of the 2007 car, with the suspension totally integrated into its aerodynamic form. In fact, the suspension arms acted as a "roof" for the lateral channels, as on the MP4-22.

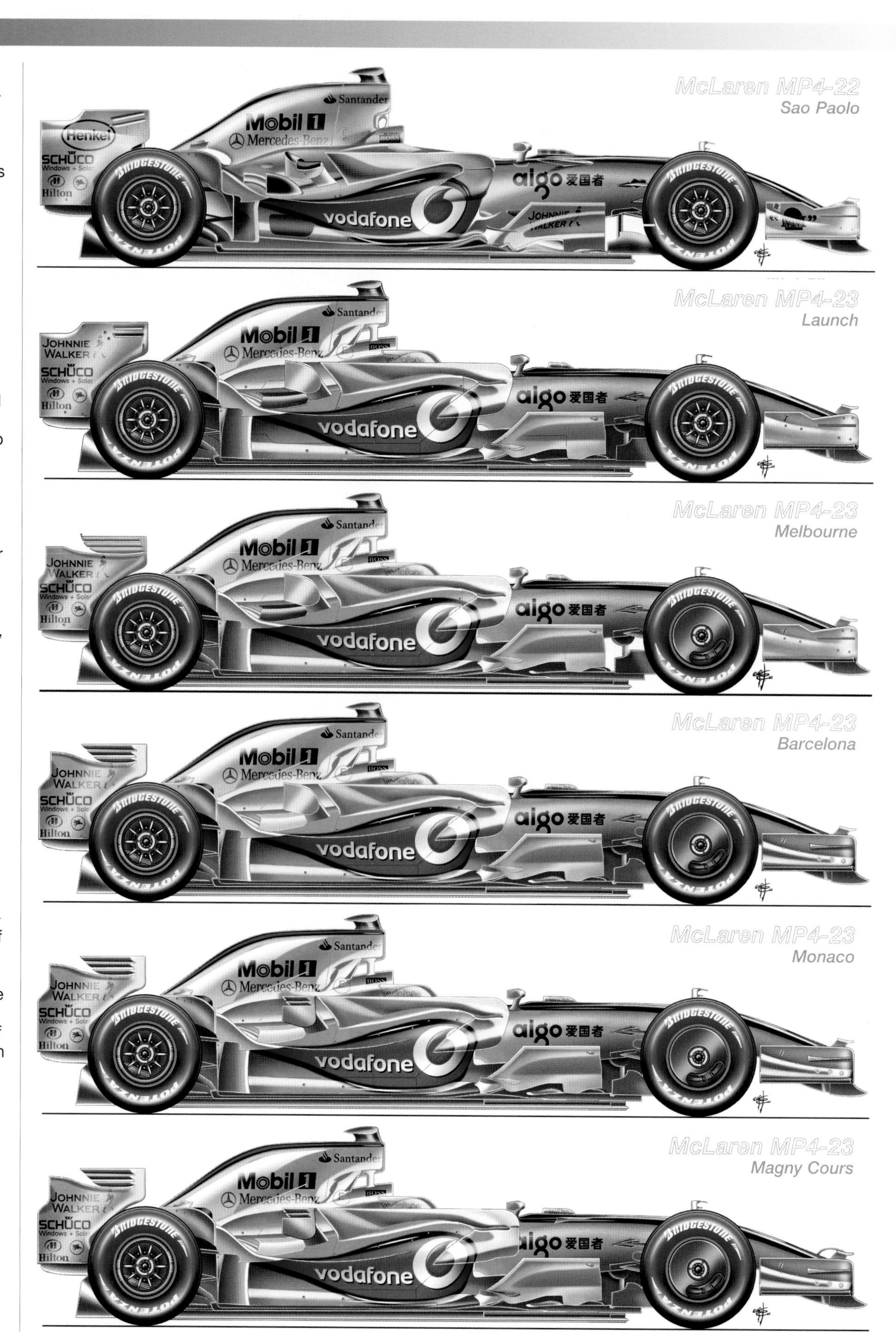

McLAREN

CONSTRUCTORS' CLASSIFICATION	2007	2008	
Position	11°	2°	+9▲
Points	0	151	+151▲

McLaren MP4-23
Silverstone

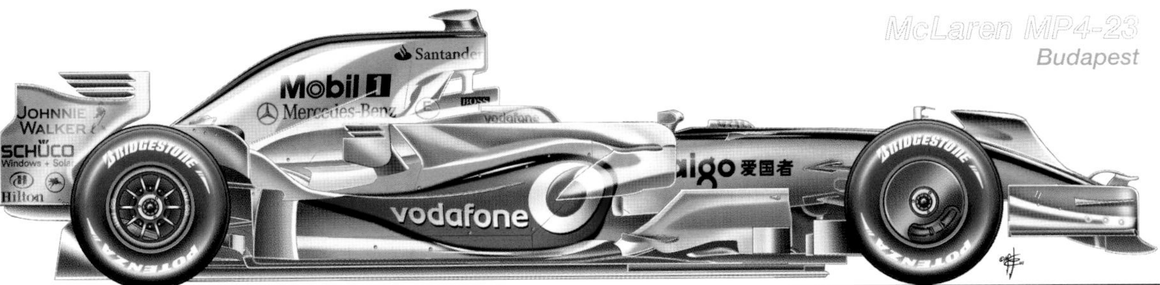

McLaren MP4-23
Budapest

McLaren MP4-23
Spa-Francorchamps

McLaren MP4-23
Monza

McLaren MP4-23
Singapore

McLaren MP4-23
Sao Paolo

Development was constant and incredible during the season, with additional aerodynamic refinements at practically every race in the battle to improve efficiency against its rival Ferrari, which set the pace throughout the season. At McLaren, there was also the tendency to try the solutions introduced by the other teams, with the result that the Anglo-German operation adopted large gullwings for the nose, first brought in by Honda. In its toe-to-toe battle with Ferrari, McLaren appeared to have slightly inferior aerodynamics, but it showed a clear superiority in the exploitation of the suspension, especially when driving over kerbs, partly due to the use of the tried and tested inertial dampers. The fairly stiff set-up of the front end and the softer rear ensured greater precision on insertion and better traction with a tendency to oversteer, which Hamilton liked. The McLarens often had a net advantage in qualifying due to their ability to immediately bring their tyres up to operational temperatures; a quality that translated into greater wear than those of the Ferraris when racing, with Hamilton sometimes having problems with the fronts and Kovalainen with the rears. To McLaren goes the kudos of achieving a further sophistication in car management: it comprised the addition of a couple of paddles behind the steering wheel, which enabled the team to partly bypass the ban on using traction control (see the Cockpits chapter). In the fight for the world drivers' title, the greater reliability of the McLaren-Mercedes-Benz was key, with just one engine breakdown experienced by Kovalainen in Japan – the team's first since 2006.

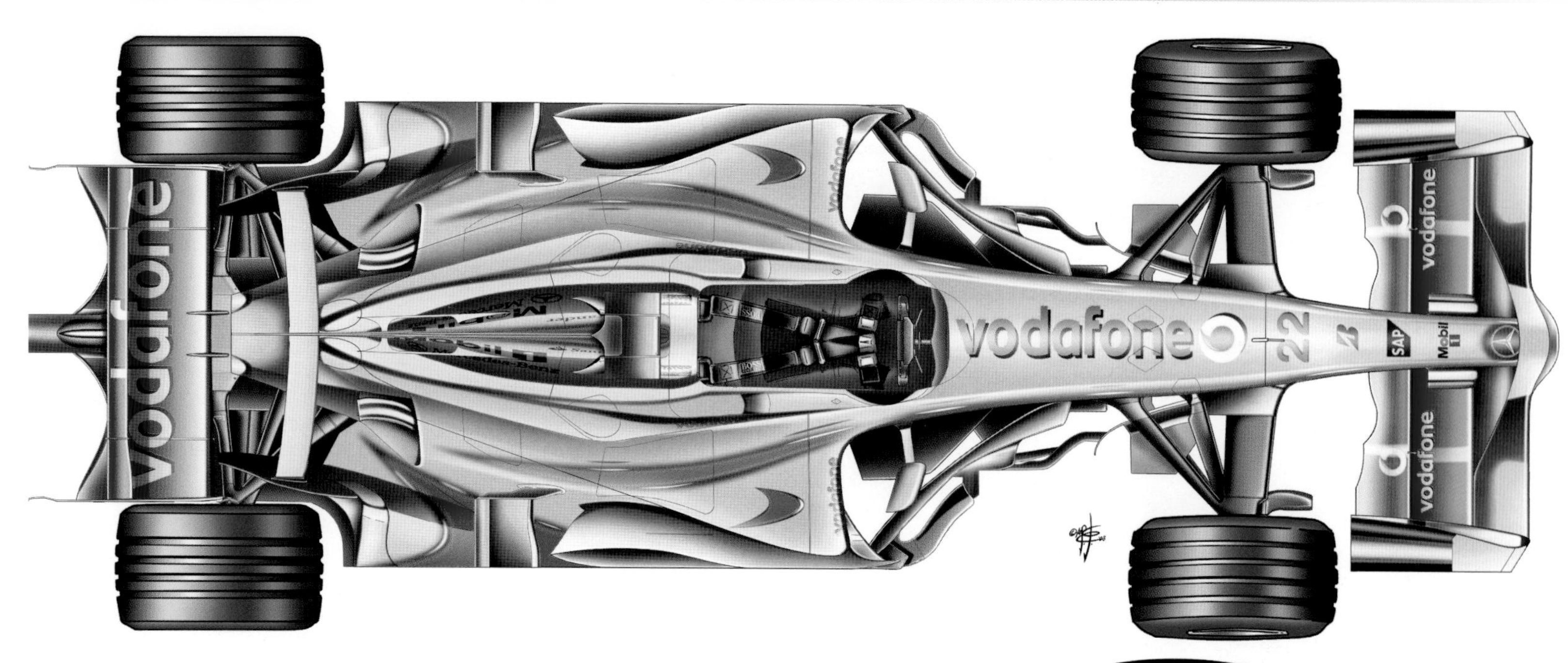

OVERHEAD VIEW

Despite the slightly longer wheelbase, in the overhead view the MP4-23 retained a family look with respect to the earlier MP4-22, above all in the shape of the nose and the second-skin-like fairing of the mechanical organs at the rear. The front wing exploited the full chord possible for each of the 3 profiles which were then topped by a bridging flap; the car was subjected to constant development in this area, with 4 profiles and Honda-style elephant's ears eventually being presented. The boomerangs in front of the sidepods formed a single element with the chimneys in order to clean up the air flow towards the rear wing.

DYNAMIC AIR INTAKE

With the Viking's horns having been discarded at the Nürburgring in 2007 (in the circle), the area of the dynamic engine air intake and the roll-bar was extensively studied in the wind tunnel with the aim of developing it into an aerodynamic element capable of cleaning up the air flow towards the rear wing; this feature required considerable structural calculations in order to ensure it passed the extremely severe crash tests to which this area is subjected.

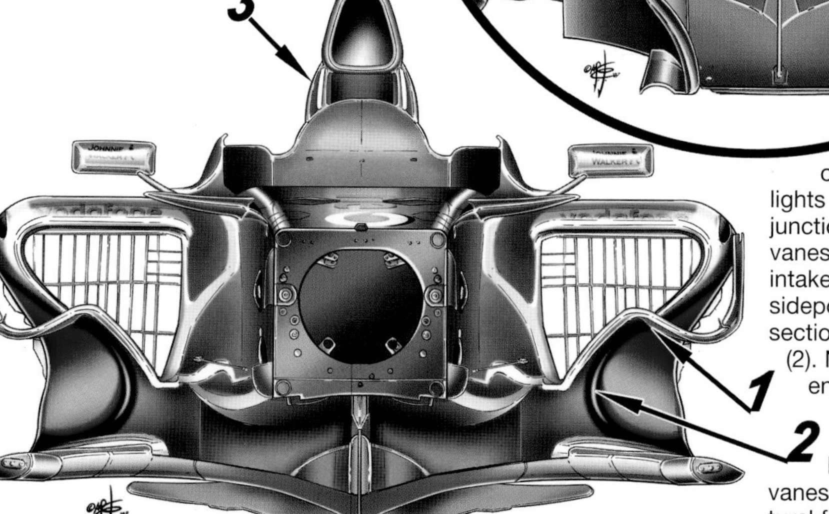

FRONT VIEW

The frontal section of the MP4-23 chassis highlights the more sophisticated junction (1) with the turning vanes and the triangular-shaped intake mouths of the MP4-22 sidepods, even though the lower section was even more tapered (2). Note how the dynamic engine air intake was smaller and separated from the area of the driver's head by the vertical turning vanes (3) that also had a structural function.

6-LEVER STEERING WHEEL

Among the most interesting developments from the season was the further refinement of the steering wheel management system introduced by McLaren with no less than 6 levers being fitted; this feature was revealed by the TV images at Monaco when the steering wheel appeared above the bodywork in the tight hairpin. The first two (1) continued to control the gear changes: up on the right and down on the left. A new pair of levers (2) was added to facilitate the driver's management of other functions such as the mapping of the engine or the differential without him having to take his hands off the wheel to adjust the more traditional knobs. Naturally, the clutch levers (3) remained unchanged in the usual position at the bottom.

MELBOURNE

The search for more downforce on the front end without altering the flow of air towards the centre of the car led to a new bridging flap, the central section of which was sufficiently neutral to better direct the air flow, while the peripheral portions close to the end-plates were devoted to creating more aerodynamic load, both through a greater angle of incidence of the profile itself and through the introduction of a vent similar to that used by Williams two years ago.

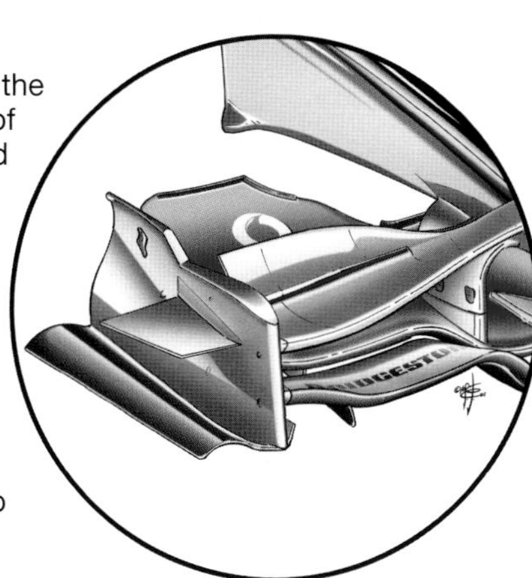

SEPANG

For the first time McLaren used a drum that completely covered the external part of the disc too. This feature was adopted together with Ferrari-style wheel covers designed to better channel the air towards the lower pressure zone outside the wheels.

REAR WING

A new rear wing for McLaren that recalled the one used by Renault in the shape of the two-piece end-plates; this configuration was designed to reduce turbulence and therefore improve penetration. Note that only in Melbourne did the team mount the unusual U-shaped profile above the rear light that had never even been track tested before.

ISTANBUL

A new front wing for McLaren that differed above all in the area of the end-plates, which presented slightly upturned external fins and a differently shaped end section that also incorporated two small Gurney flaps (one vertical and one horizontal). In the comparison with the wing from the previous season note the direct correspondence of the shapes of both the central section (less deep), the three dished profiles and the end-plates aligned towards the inside of the wheels.

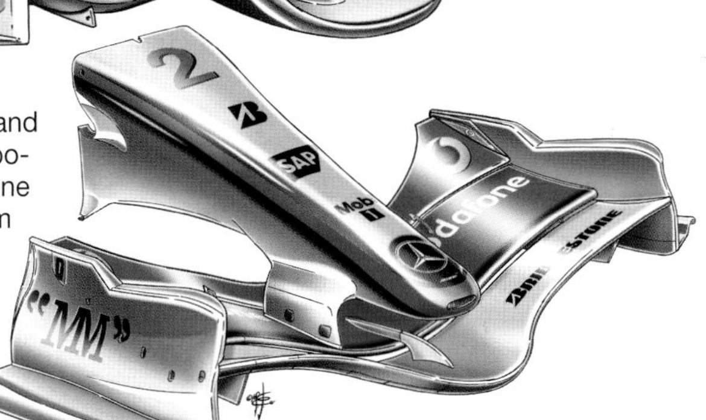

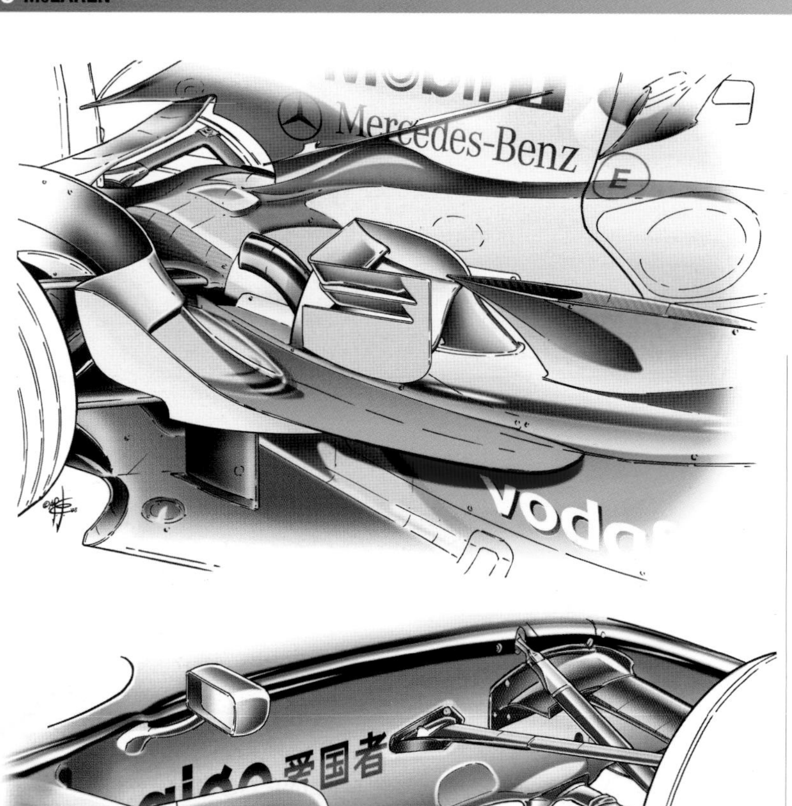

MONACO

New Renault-style vented end-plates to increase the efficiency of the winglets behind the chimneys on the McLaren, as well as doubled-up turning vanes immediately behind the front wheels. The Renault-style vents in the end-plates allowed a reduction in the negative effects of the vortices created in this area.

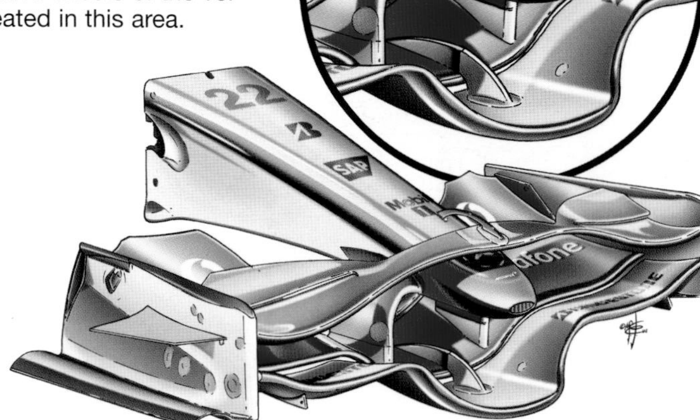

MONTREAL: FRONT WING

The high speed characteristics of the Montreal circuit require a specific aerodynamic configuration. McLaren therefore fielded an ad hoc package, starting with the front wing, which as well as having different profiles above all featured the strengthening bridge (enlarged in the circle) required by the FIA to overcome the controversial flexing.

TURNING VANES

The turning vanes were modified, with the doubling up of the small end-plates to better manage the turbulence generated by the front wheels.

ENGINE COVER

The engine cover was also new and more tapered towards the rear. It again featured end-plates on the flaps with no vents.

REAR WING

The rear wing for medium-fast tracks was also new with less aggressive profiles arching upwards in the centre.
Note the presence of a winglet between the vertical pylons and the absence of the U-shaped appendix above the deformable structure, where instead the directional fin remained.

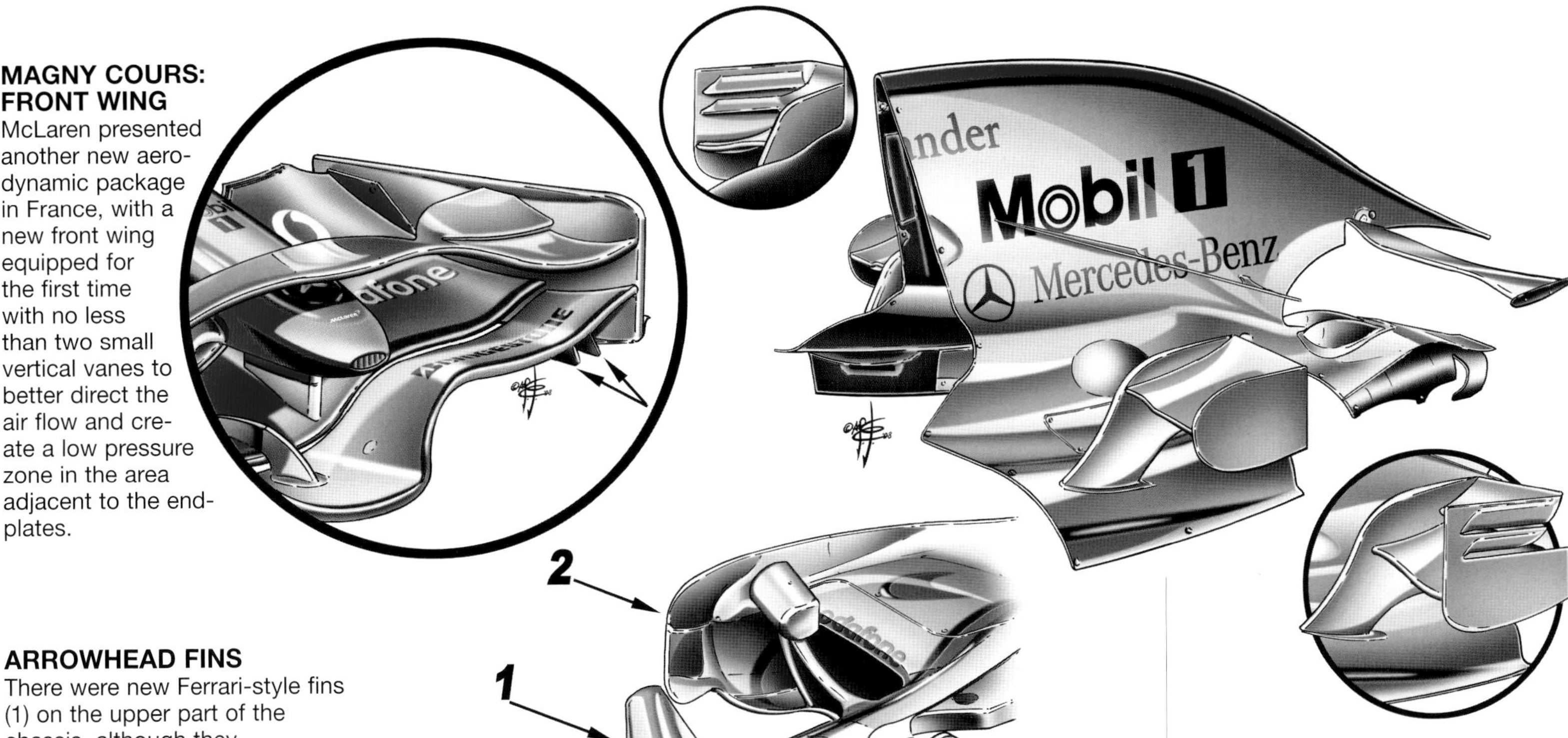

MAGNY COURS: FRONT WING

McLaren presented another new aerodynamic package in France, with a new front wing equipped for the first time with no less than two small vertical vanes to better direct the air flow and create a low pressure zone in the area adjacent to the end-plates.

ARROWHEAD FINS

There were new Ferrari-style fins (1) on the upper part of the chassis, although they had less of an arrowhead shape in order to realign the air flow. The lateral boomerangs became longer to make the turning vane zone more efficient and create more downforce (around 14% of the total loading).

ENGINE COVER

The new engine cover introduced in Canada was retained, but it was combined with the high loading Monaco winglets equipped with Renault-style vented end-plates.

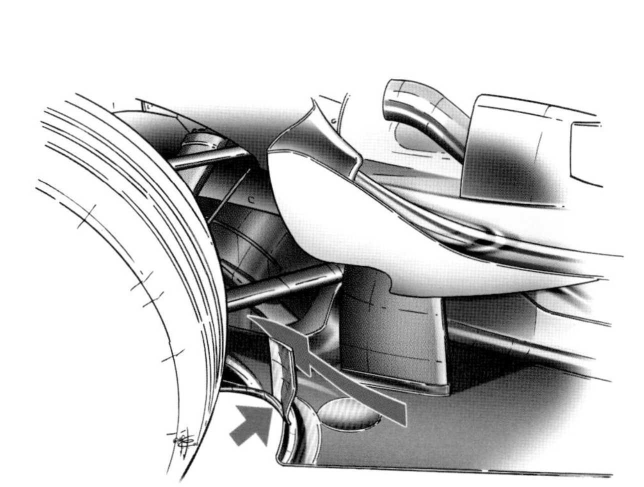

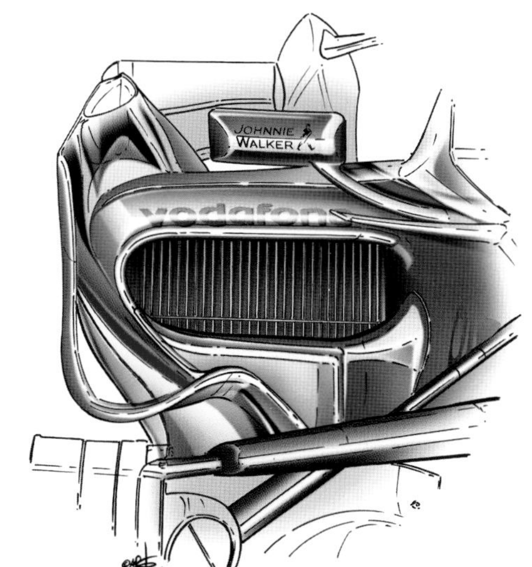

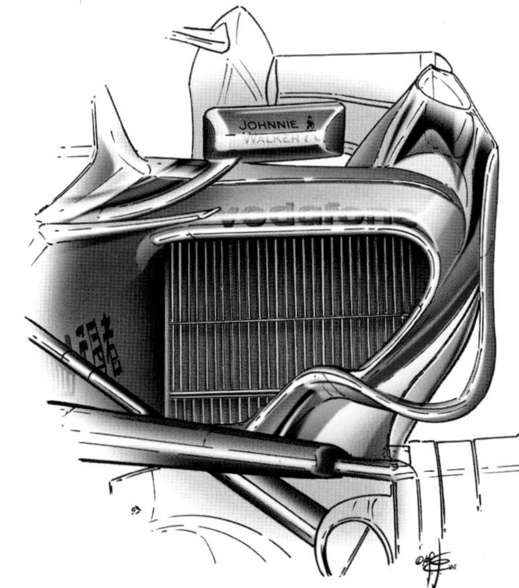

SILVERSTONE

Doubling up of the vertical vanes in the area ahead of the rear wheels on the McLaren, with the aim of better channelling the air inside the wheels rather than outside as the flow would tend to do in this area.

ASYMMETRIC SIDEPODS

While this was not a brand new feature, intakes so notably asymmetric as those introduced on the MP4-23 at Silverstone where the ambient temperatures were relatively low had never been seen before.
The right-hand pod carried the water radiators with the intake mouth being significantly reduced to the benefit of aerodynamic efficiency.
This feature was used again at Spa.

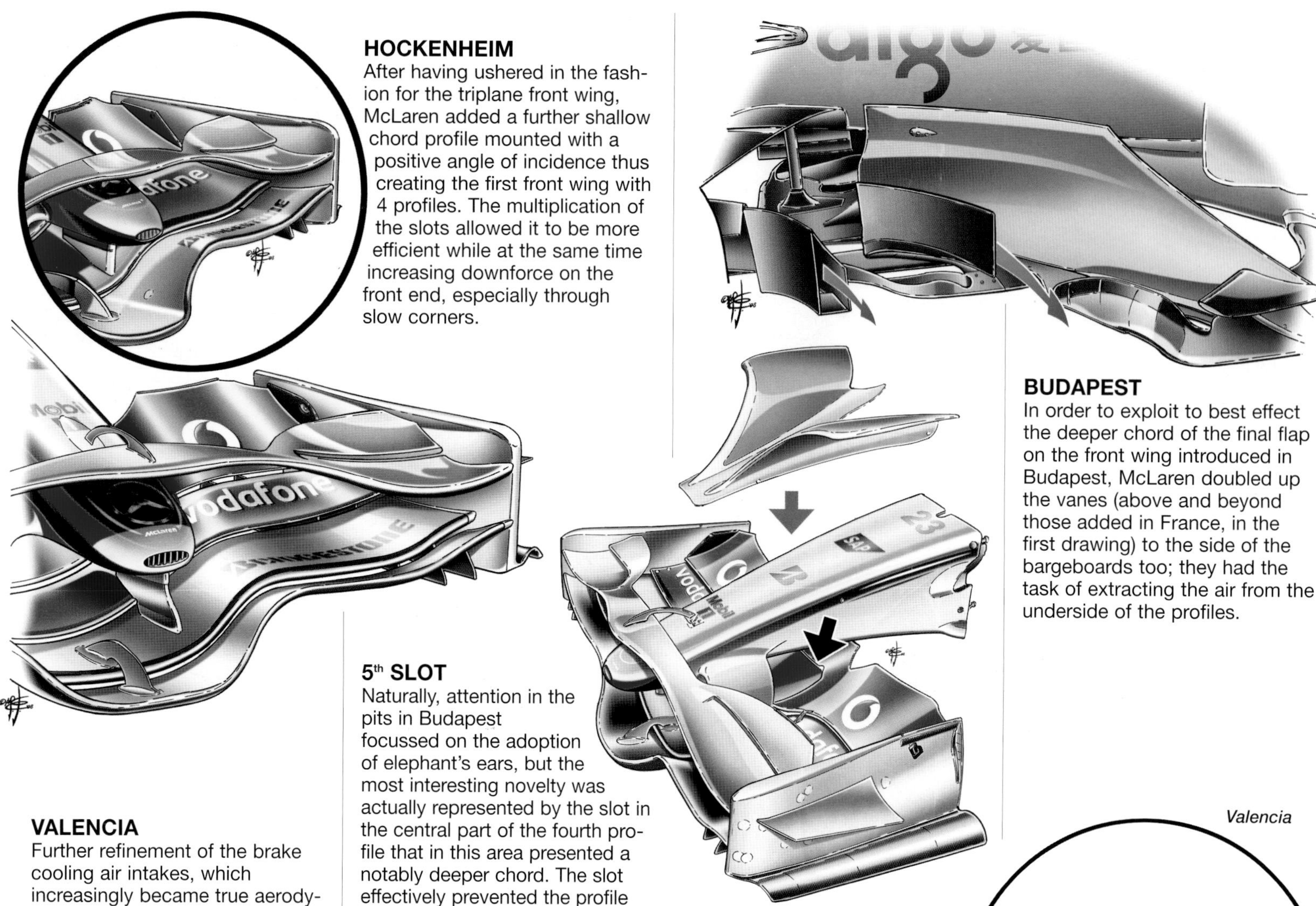

HOCKENHEIM

After having ushered in the fashion for the triplane front wing, McLaren added a further shallow chord profile mounted with a positive angle of incidence thus creating the first front wing with 4 profiles. The multiplication of the slots allowed it to be more efficient while at the same time increasing downforce on the front end, especially through slow corners.

BUDAPEST

In order to exploit to best effect the deeper chord of the final flap on the front wing introduced in Budapest, McLaren doubled up the vanes (above and beyond those added in France, in the first drawing) to the side of the bargeboards too; they had the task of extracting the air from the underside of the profiles.

5th SLOT

Naturally, attention in the pits in Budapest focussed on the adoption of elephant's ears, but the most interesting novelty was actually represented by the slot in the central part of the fourth profile that in this area presented a notably deeper chord. The slot effectively prevented the profile from stalling at certain angles of incidence. With regard to the feature copied from Honda, McLaren only passed the crash test at the last moment on Friday morning.

VALENCIA

Further refinement of the brake cooling air intakes, which increasingly became true aerodynamic devices. In Valencia, McLaren added this further fin to facilitate extraction of the air from the underside of the car therefore increase downforce acting on the rear end.

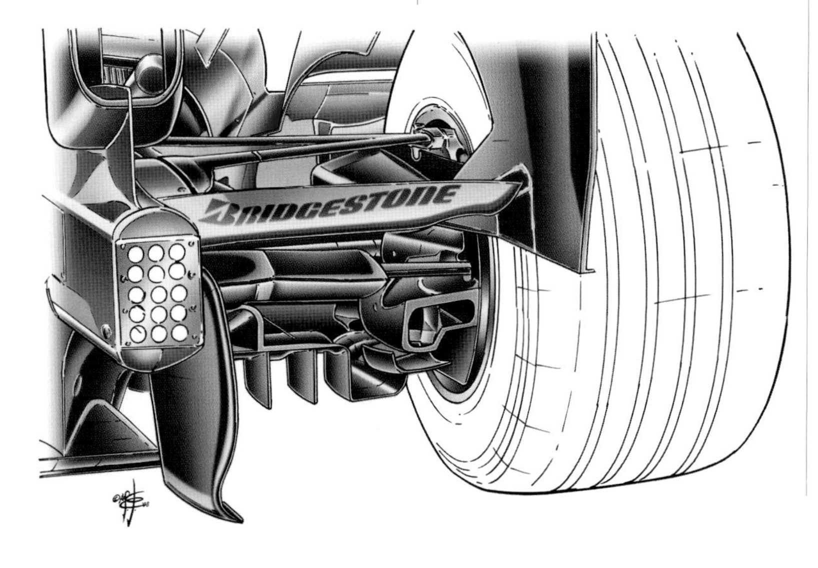

Valencia

SPA

On the fast Spa circuit, the MP4-23s retained the gull-wing fins introduced in Hungary and used again in Valencia, combining them this time with a front wing with just 3 profiles (compared in the drawing with the 4-profile version). This choice was perhaps made in order to exploit stalling in the fast corners.

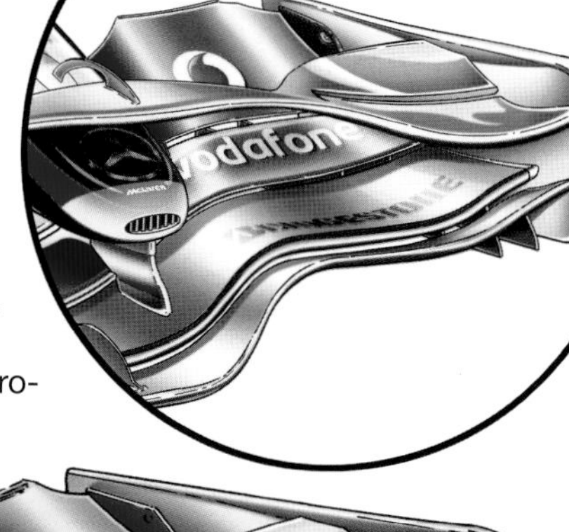

Spa-Francorchamps

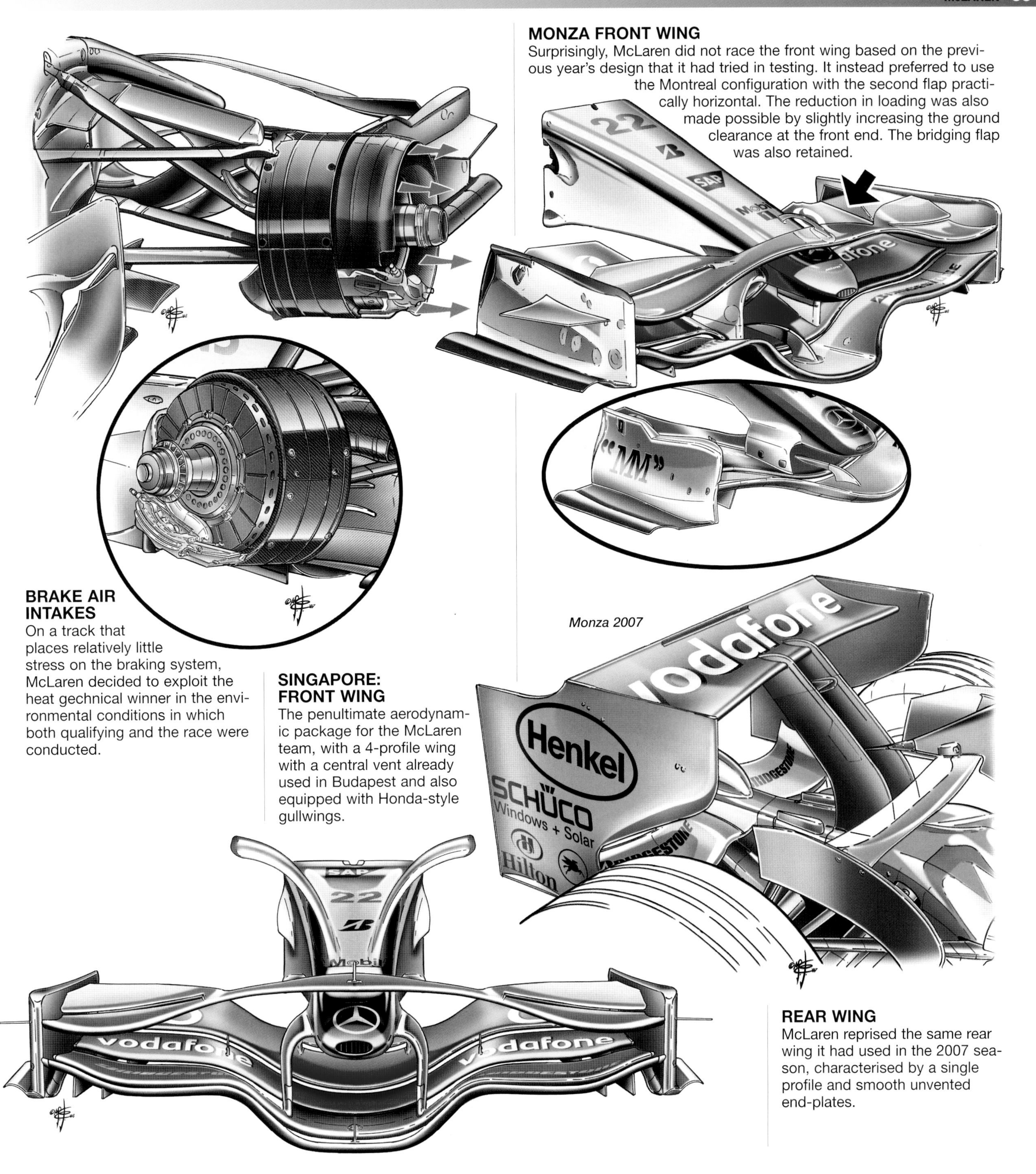

Monza 2007

MONZA FRONT WING

Surprisingly, McLaren did not race the front wing based on the previous year's design that it had tried in testing. It instead preferred to use the Montreal configuration with the second flap practically horizontal. The reduction in loading was also made possible by slightly increasing the ground clearance at the front end. The bridging flap was also retained.

BRAKE AIR INTAKES

On a track that places relatively little stress on the braking system, McLaren decided to exploit the heat gechnical winner in the environmental conditions in which both qualifying and the race were conducted.

SINGAPORE: FRONT WING

The penultimate aerodynamic package for the McLaren team, with a 4-profile wing with a central vent already used in Budapest and also equipped with Honda-style gullwings.

REAR WING

McLaren reprised the same rear wing it had used in the 2007 season, characterised by a single profile and smooth unvented end-plates.

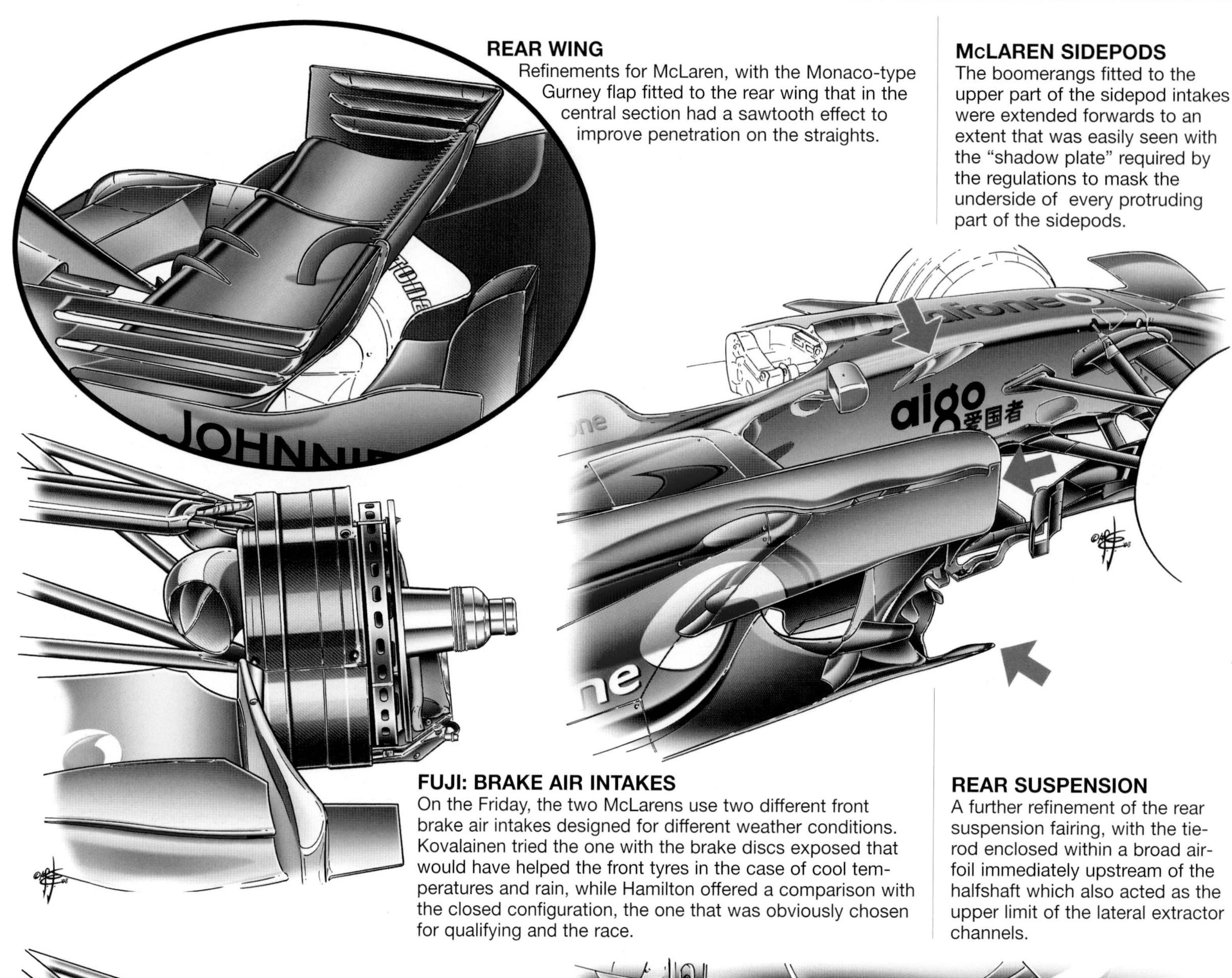

REAR WING
Refinements for McLaren, with the Monaco-type Gurney flap fitted to the rear wing that in the central section had a sawtooth effect to improve penetration on the straights.

McLAREN SIDEPODS
The boomerangs fitted to the upper part of the sidepod intakes were extended forwards to an extent that was easily seen with the "shadow plate" required by the regulations to mask the underside of every protruding part of the sidepods.

FUJI: BRAKE AIR INTAKES
On the Friday, the two McLarens use two different front brake air intakes designed for different weather conditions. Kovalainen tried the one with the brake discs exposed that would have helped the front tyres in the case of cool temperatures and rain, while Hamilton offered a comparison with the closed configuration, the one that was obviously chosen for qualifying and the race.

REAR SUSPENSION
A further refinement of the rear suspension fairing, with the tie-rod enclosed within a broad airfoil immediately upstream of the halfshaft which also acted as the upper limit of the lateral extractor channels.

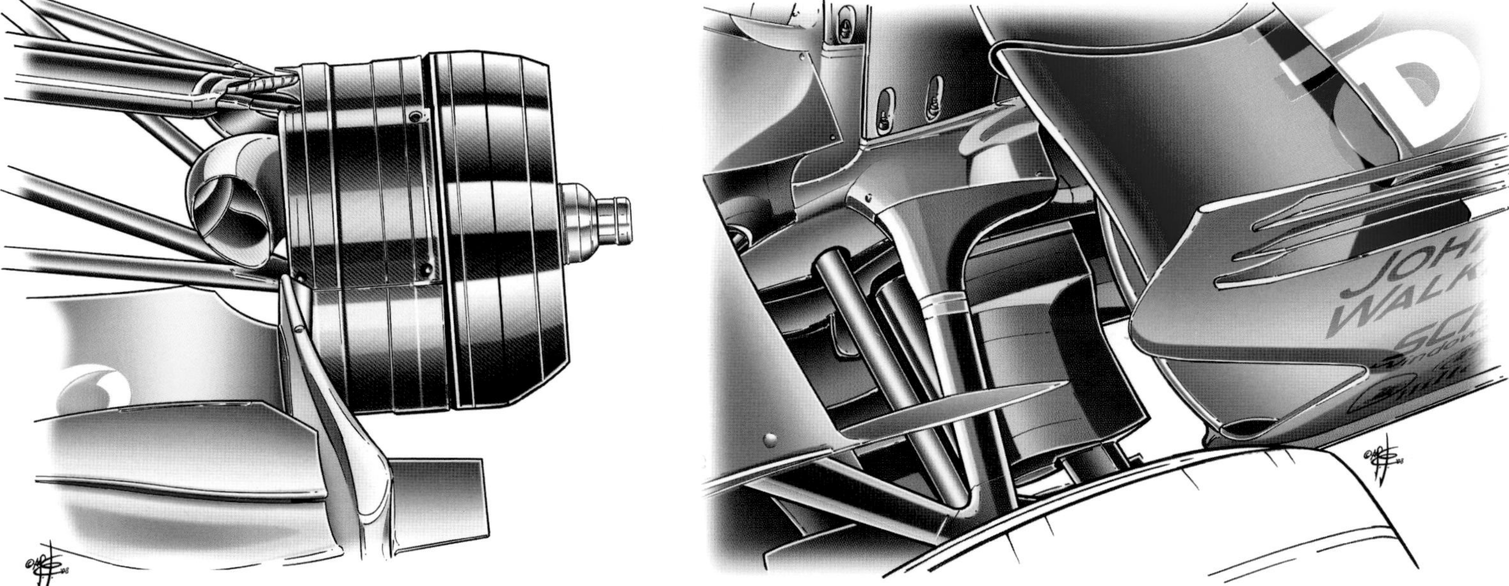

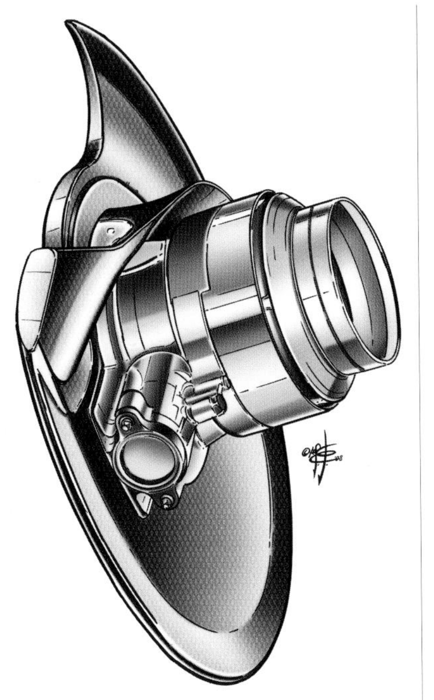

WHEEL COVER FIXTURE

McLaren further developed the attachment of the wheel cover in a pre-established fixed position with a system equipped with a small piston actuated by the extraction of the air wrench, thus speeding up the replacement procedure.

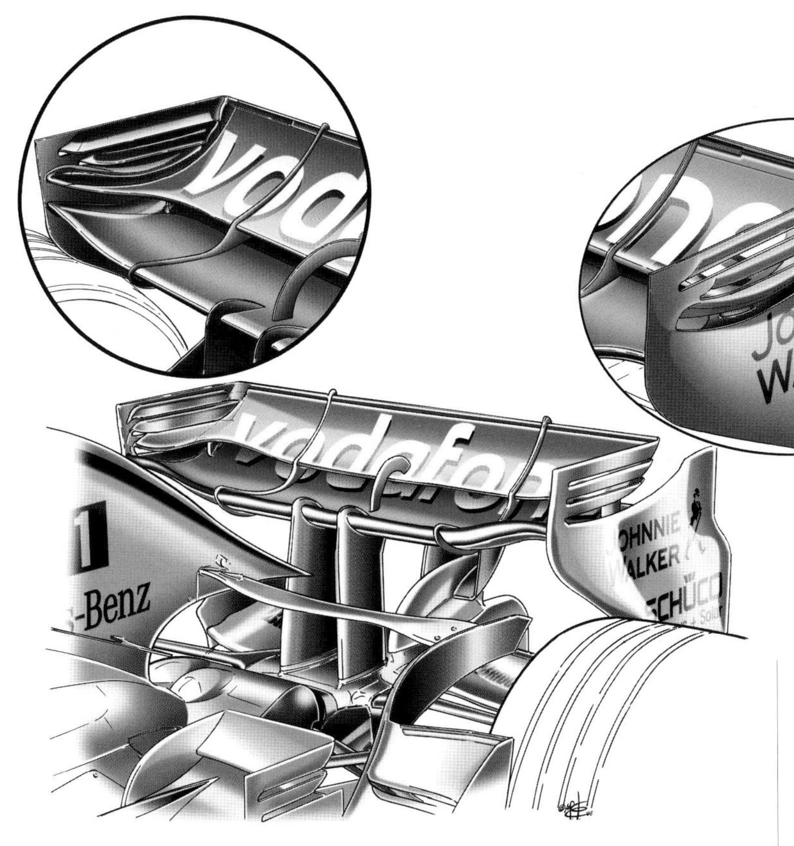

SAO PAOLO

The last race of the season and a new aerodynamic package for McLaren, announced by Martin Withmarsh in Singapore on the occasion of the penultimate development of the MP4-23. The easiest element to identify is the rear wing representing a development of the previous version depicted in the drawing. The modifications were designed to reduce the turbulence generated by the trailing edge and the end-plates that were subjected to painstaking development. The mounting of the flap with the end-plate extending forwards in the form of a finger, already present on the previous version, was emphasised, as can be seen in the comparison (the new configuration shown in the circle).

MP4-24 SAO PAOLO

A CAD view of the MP4-23 that carried Lewis Hamilton to the World Championship, created on the basis of the photos gathered during the course of the season; an excellent medium for television that offers the possibility of illustrating the new features almost in real time.

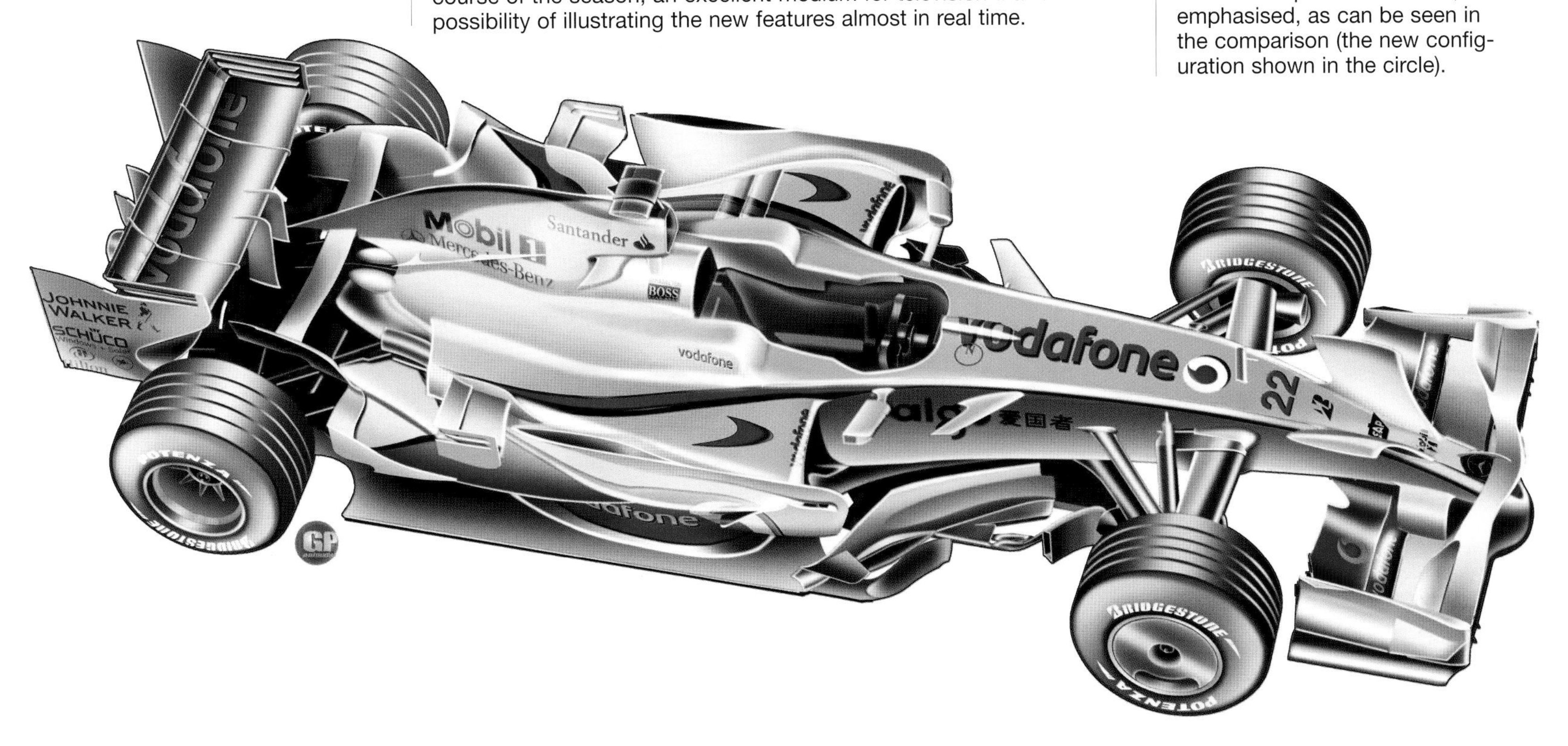

After a second place in the 2007 world championship by arbitration following the disqualification of McLaren, in 2008 BMW became part of the small group of cars able to fight for victory.
The new project saw massive use of the wind tunnel, especially in correlation with CFD analysis, to verify the aerodynamic efficiency of the F1 08.
Yet despite the notable progress achieved over the F1 07, the new car was still unable to bridge the gap between itself and the top two of the class, Ferrari and McLaren, even though the team's start of the season was promising.
The BMW seemed slightly late in its development programme in relation to the two great protagonists of the season. Kubica's victory in Canada, favoured by the mess-up of Hamilton and Rosberg in the pits, was just recompense for the substantial efforts of the team directed by Mario Thiessen, even if new solutions introduced were often not confirmed as beneficial on the track, development taking place in alternate phases.
In spite of all this, BMW can still boast of having brought in new aerodynamic features that led the way and which will be amply analysed in the Aerodynamics chapter.
As far as the F1 08 is concerned, the elimination of traction control meant the construction of a more stable and 'sincere' car at the front, with a rear end able to produce a great deal of grip. In other words, it was easier to drive and more predictable.
The objective of this extensive aerodynamic research carried out during the design stage of the new car was to achieve a major improvement over the F1 07.
The choice of the group directed by Willy Rampf was based on a car with a slightly longer wheelbase, made possible by distancing the front axle in relation to the car's body.
That gave more space and freedom for the positioning of the turning vanes, a very aggressive front wing able to generate considerable downforce in the central zone, which married best with a discreet range of variation of the

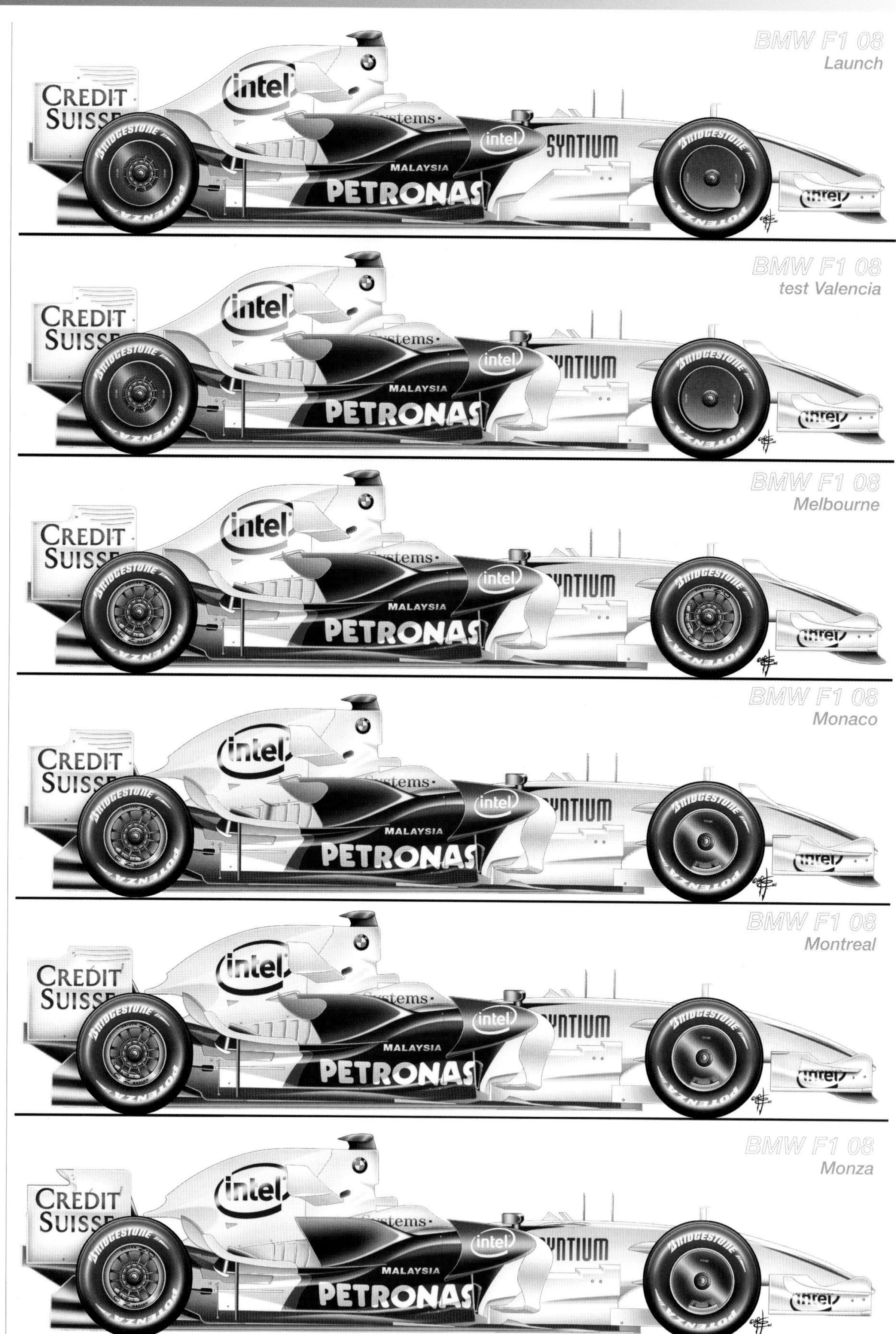

CONSTRUCTORS' CLASSIFICATION	2006	2007	
Position	2°	3°	-1 ▼
Points	101	135	+34 ▲

distribution of weights; an engine cover that also guaranteed good stability; and a good cooling system that did not penalise the aerodynamics.
The choice in this latter sector fell on a combination of the chimneys and bladed louvers.
At the static launch, the F1 08 also had covers on the front wheels, the official use of which was postponed until the Grand Prix of Monaco, while there were no 'Tomcat' fins and the vertical turning vanes in front of the sidepods that were adopted immediately at the first shakedown in Valencia, remaining on the car for its debut in Australia.
Both of these features made converts: the turning vanes were copied by almost all the other teams, which also happened with the 'Tomcat' fins of the McLaren, even if in a different shape.
The former cleaned harmful turbulence in the air flow in the delicate area behind the front wheels, so improving the efficiency of the underbody and the extractor channels.
The latter improved the quality of the air flow that hit the central part of the car, favouring the efficiency of the rear wing.
Despite a considerable amount of development, especially concentrated on the front and rear wings, BMW was never able to eliminate the technical gap between them and the two top cars, Ferrari and McLaren.
In fact, after the victory in Canada there was less progress and on a number of occasions new developments taken to the circuits did not turn out to be better than the previous solutions, or needed a long period of adaptation.
One example was the fairing of the front wheels, which was already on the car at its launch and the first test in Valencia.
But it was only adopted from the Grand Prix of Monaco and in a different shape.
Their use considerably influenced the aerodynamic balance of the car.

SIDE VIEWS
At the launch of the car, the F1 08 was shown with wheel fairing, new integrated flaps at the chimneys but without the 'Tomcat' nose and the new turning vanes in front of the sidepods, introduced in the very first tests at Valencia. At its seasonal debut in Melbourne, the front wheel fairing had disappeared as it needed a long period of setting up. At Monaco, we saw a high downforce set-up with new front and rear wings and an added flap immediately behind the chimneys, which reappeared in Budapest and Singapore. The front wheel covers were no longer on the car. Immediately after Montreal with a low downforce set-up we saw the new front wing without the 'Tomcat' fins and that rear wing. At Monza, BMW fielded an extreme car: it had a new nose and bi-plane wings, no flap on the sidepods and an extremely small rear wing. Then, for the rest of the season, it returned to a configuration similar to the one in Spain, even if with modifications of detail at every race.

TOP VIEW
In these two designs, only the static presentation version of the car and the one that made its debut in Melbourne are compared. The latter with the arrival of the nose with the new 'Tomcat' fins and, more important, with vertical turning vanes connecting with the lower part and the upper in front of the sidepod entrance, which was copied by almost all the teams.

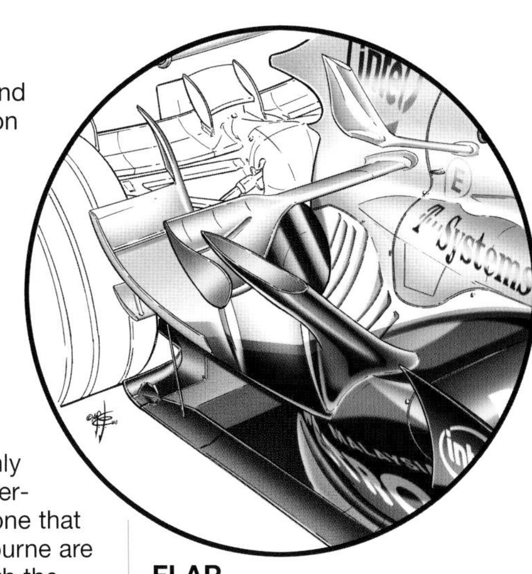

FLAP
The advancement of the 100 cm wide plane was a new and interesting feature, as it was usually placed at the height of the rear axle so as to connect it in a single element with the flap linked to the chimneys.

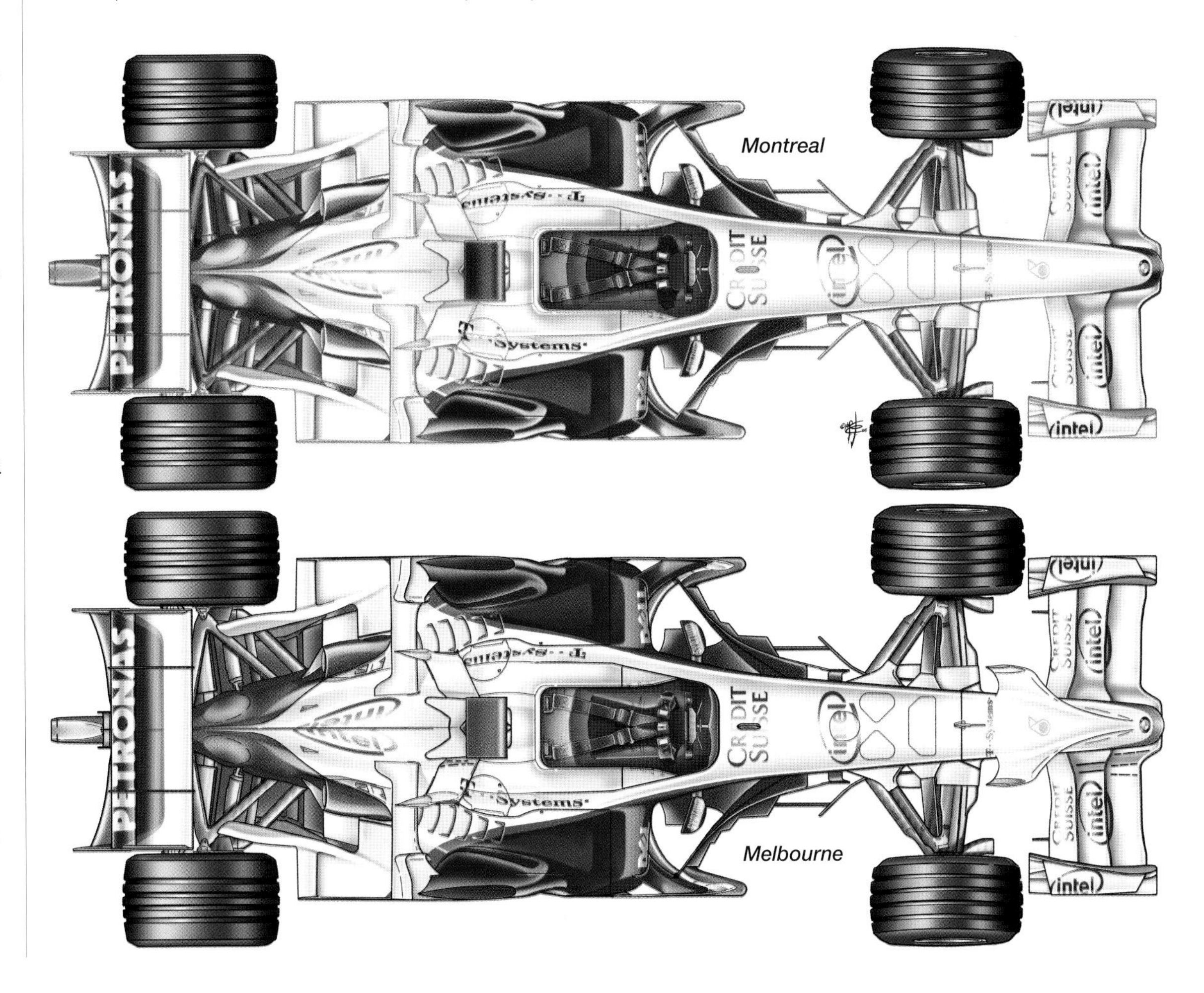

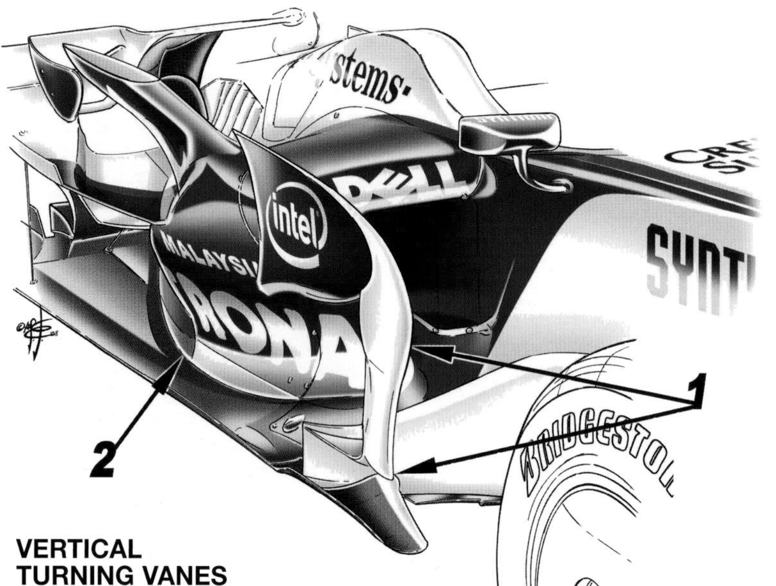

VERTICAL TURNING VANES
BMW was a lesson to the rest with this vertical connection (1) between the 'boomerangs' in the upper part of the sidepods and the turning vanes down low. Note the considerable break-up (2) of the lower part of the 'pods.

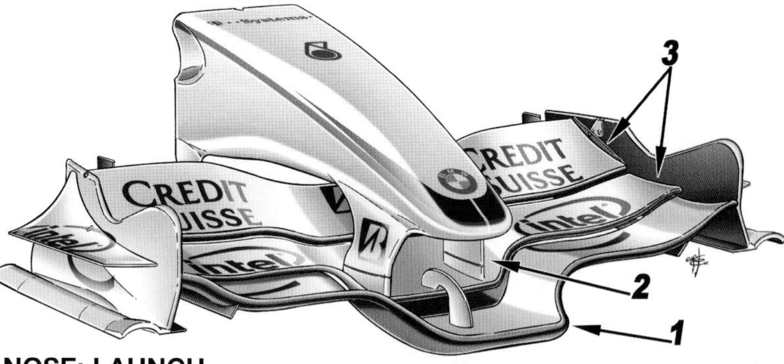

NOSE: LAUNCH
At the static launch, the nose of the F1 08 had a heavily stepped part in the centre (1) to create good vertical downforce, the central pylons (2) linked to the first flap and not the principal plane, following the example of McLaren, as with the upper deck flaps (3).

NOSE SHAKEDOWN
At the first shakedown in Valencia, only the two symmetrical planes were fitted to simulate the position of the television camera, which would remain on the nose. In the afternoon, the 'Tomcat' fins were also mounted to improve the quality of the air flow towards the body of the car.

SEPANG
BMW was the only team to introduce a number of new features as early as the second race at Sepang, where we saw two small evolutions of the fins in the area inside the front suspension. The circle indicates the solution used in Melbourne, while the design shows the new one of horn shape turned forward.

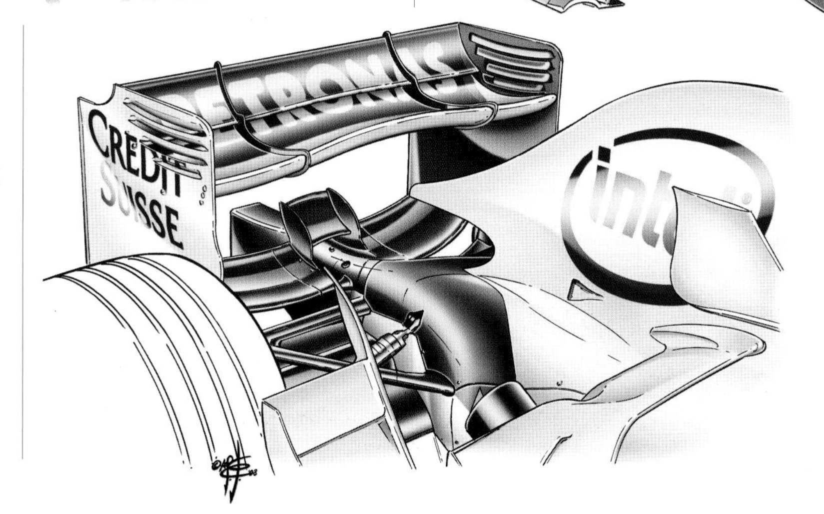

SAKHIR
For this circuit, both BMW and Ferrari chose more unladen aerodynamic set-ups: the German constructor opted for a convex plane very similar to the one used at Montreal, but with greater incidence and especially with end plates of medium-high load.

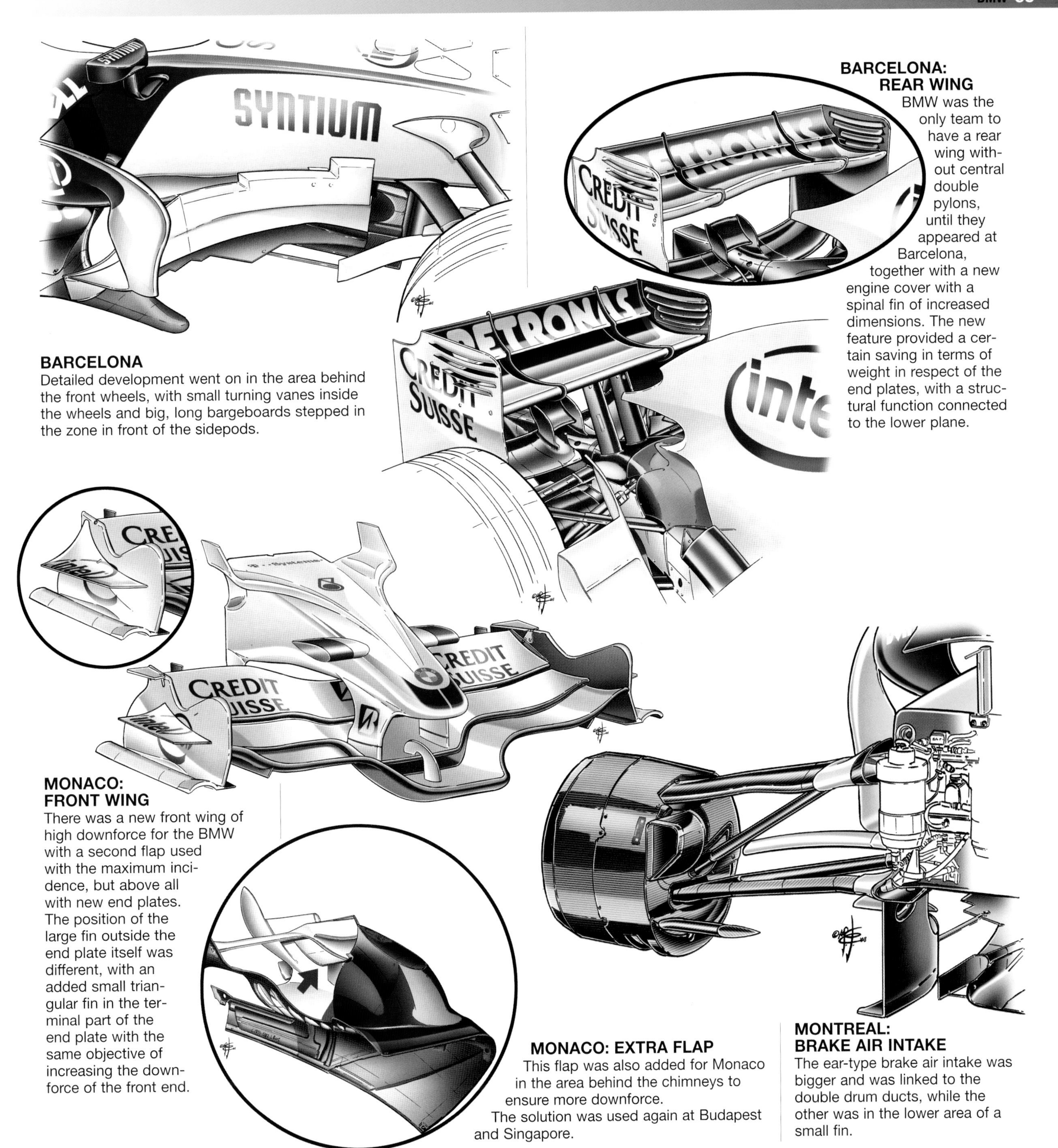

BARCELONA: REAR WING
BMW was the only team to have a rear wing without central double pylons, until they appeared at Barcelona, together with a new engine cover with a spinal fin of increased dimensions. The new feature provided a certain saving in terms of weight in respect of the end plates, with a structural function connected to the lower plane.

BARCELONA
Detailed development went on in the area behind the front wheels, with small turning vanes inside the wheels and big, long bargeboards stepped in the zone in front of the sidepods.

MONACO: FRONT WING
There was a new front wing of high downforce for the BMW with a second flap used with the maximum incidence, but above all with new end plates. The position of the large fin outside the end plate itself was different, with an added small triangular fin in the terminal part of the end plate with the same objective of increasing the downforce of the front end.

MONACO: EXTRA FLAP
This flap was also added for Monaco in the area behind the chimneys to ensure more downforce.
The solution was used again at Budapest and Singapore.

MONTREAL: BRAKE AIR INTAKE
The ear-type brake air intake was bigger and was linked to the double drum ducts, while the other was in the lower area of a small fin.

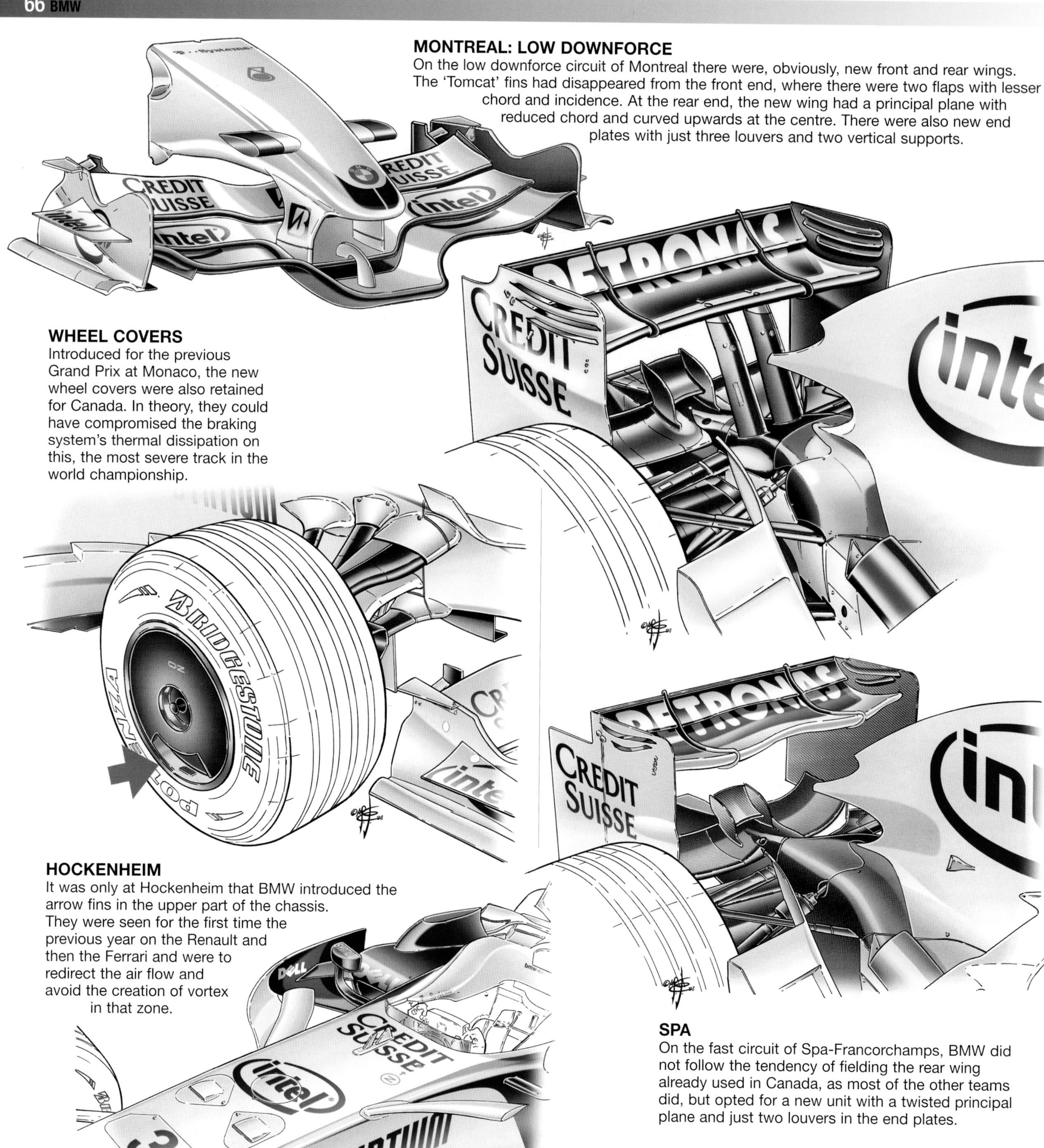

MONTREAL: LOW DOWNFORCE

On the low downforce circuit of Montreal there were, obviously, new front and rear wings. The 'Tomcat' fins had disappeared from the front end, where there were two flaps with lesser chord and incidence. At the rear end, the new wing had a principal plane with reduced chord and curved upwards at the centre. There were also new end plates with just three louvers and two vertical supports.

WHEEL COVERS

Introduced for the previous Grand Prix at Monaco, the new wheel covers were also retained for Canada. In theory, they could have compromised the braking system's thermal dissipation on this, the most severe track in the world championship.

HOCKENHEIM

It was only at Hockenheim that BMW introduced the arrow fins in the upper part of the chassis. They were seen for the first time the previous year on the Renault and then the Ferrari and were to redirect the air flow and avoid the creation of vortex in that zone.

SPA

On the fast circuit of Spa-Francorchamps, BMW did not follow the tendency of fielding the rear wing already used in Canada, as most of the other teams did, but opted for a new unit with a twisted principal plane and just two louvers in the end plates.

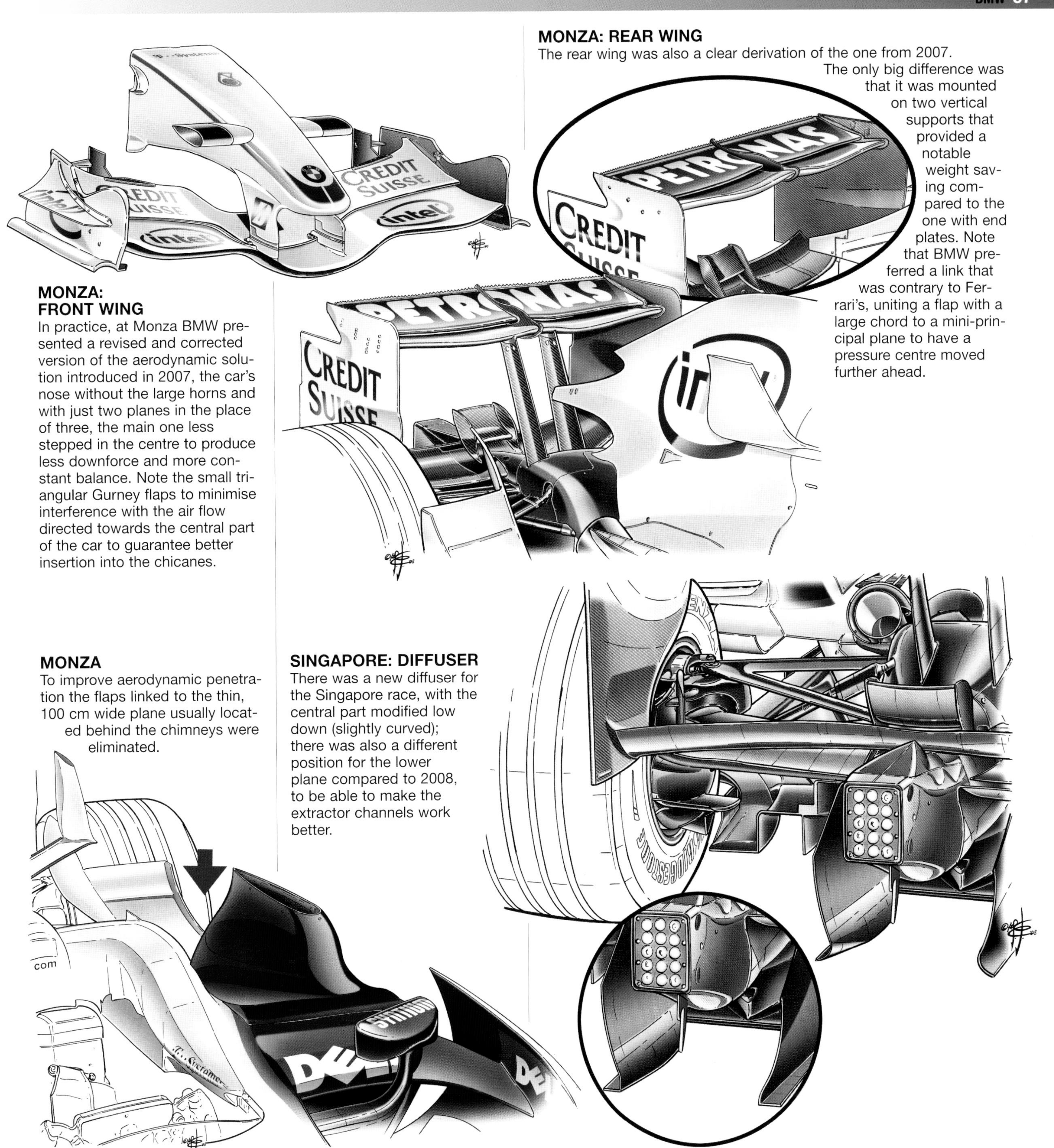

MONZA: REAR WING

The rear wing was also a clear derivation of the one from 2007. The only big difference was that it was mounted on two vertical supports that provided a notable weight saving compared to the one with end plates. Note that BMW preferred a link that was contrary to Ferrari's, uniting a flap with a large chord to a mini-principal plane to have a pressure centre moved further ahead.

MONZA: FRONT WING

In practice, at Monza BMW presented a revised and corrected version of the aerodynamic solution introduced in 2007, the car's nose without the large horns and with just two planes in the place of three, the main one less stepped in the centre to produce less downforce and more constant balance. Note the small triangular Gurney flaps to minimise interference with the air flow directed towards the central part of the car to guarantee better insertion into the chicanes.

MONZA

To improve aerodynamic penetration the flaps linked to the thin, 100 cm wide plane usually located behind the chimneys were eliminated.

SINGAPORE: DIFFUSER

There was a new diffuser for the Singapore race, with the central part modified low down (slightly curved); there was also a different position for the lower plane compared to 2008, to be able to make the extractor channels work better.

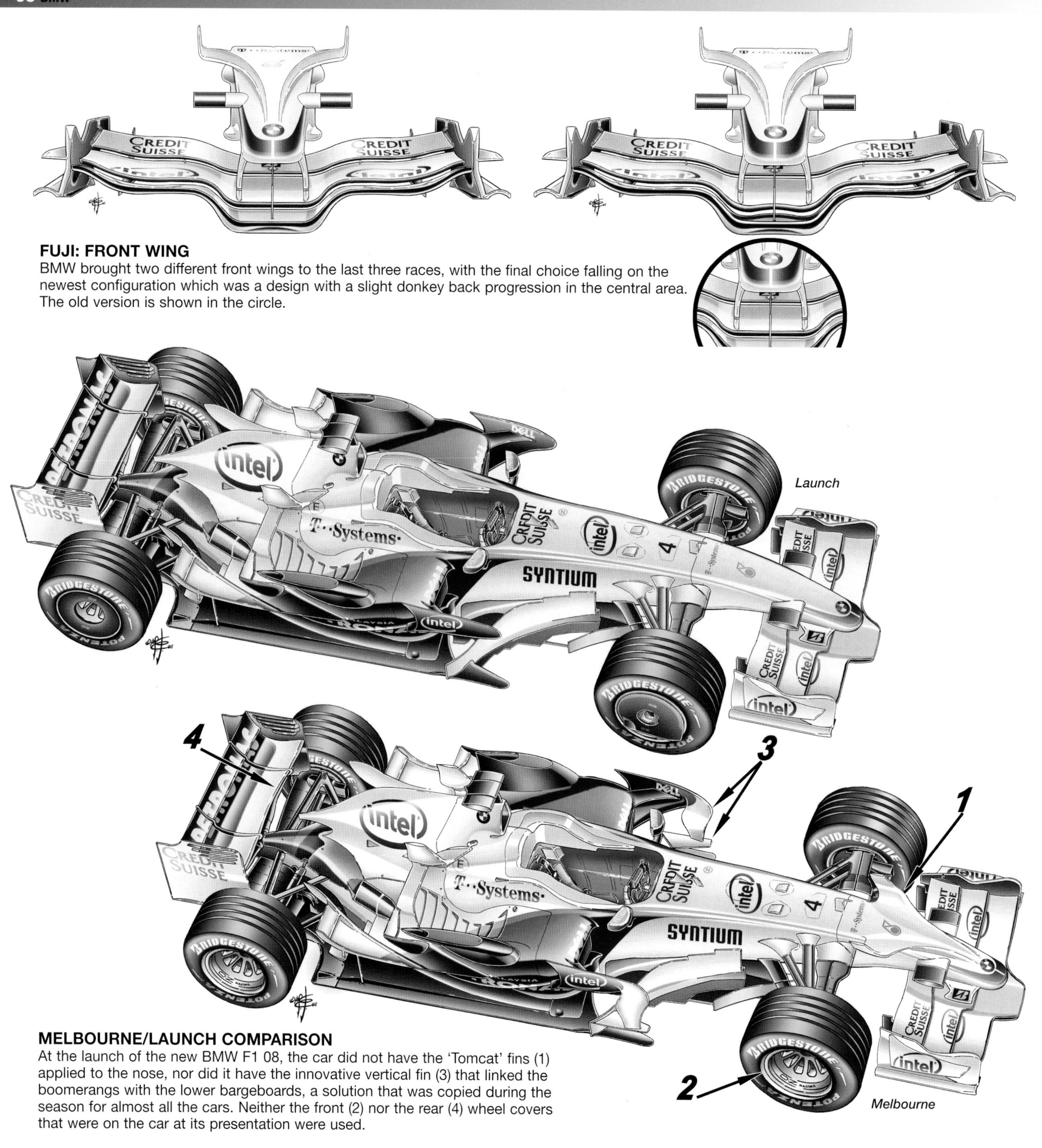

FUJI: FRONT WING

BMW brought two different front wings to the last three races, with the final choice falling on the newest configuration which was a design with a slight donkey back progression in the central area. The old version is shown in the circle.

MELBOURNE/LAUNCH COMPARISON

At the launch of the new BMW F1 08, the car did not have the 'Tomcat' fins (1) applied to the nose, nor did it have the innovative vertical fin (3) that linked the boomerangs with the lower bargeboards, a solution that was copied during the season for almost all the cars. Neither the front (2) nor the rear (4) wheel covers that were on the car at its presentation were used.

MONACO

A high downforce package was introduced for the especially slow Monaco circuit. As well as the planes with a larger chord, the front wings had new end plates (1) with a small triangular fin (2) in the area close to the wheels. The wheel covers (3) appeared, but only for the front end. To increase load, flaps (4) were added behind the chimneys. The new wings with vertical supports (5) that first appeared in Spain were used with bigger incidence planes (6).

MONTREAL

The BMW had a low downforce set-up for Montreal with various end plates for the front wing (1), a reduced chord flap (2), a nose without 'Tomcat' fins (3) and a new rear wing in both the principal plane with a reduced chord (4) and end plates.

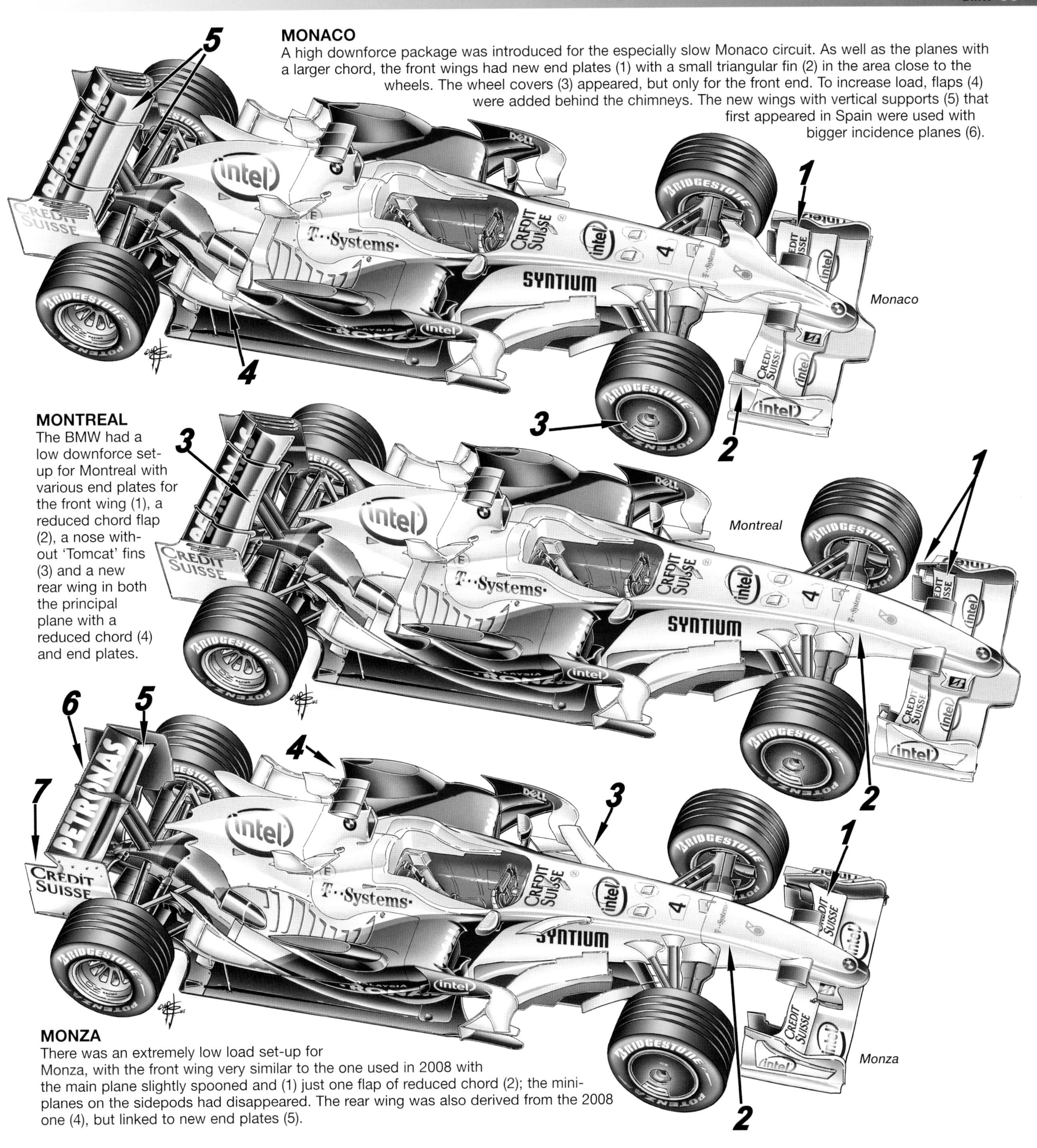

MONZA

There was an extremely low load set-up for Monza, with the front wing very similar to the one used in 2008 with the main plane slightly spooned and (1) just one flap of reduced chord (2); the mini-planes on the sidepods had disappeared. The rear wing was also derived from the 2008 one (4), but linked to new end plates (5).

In practice, Renault maintained the same position as in the 2007 season, given that its 3rd place that year was due to the retrocession of McLaren to last. But Fernando Alonso scored more points with his win in the GP of Singapore and 4th place in the subsequent GP of Japan. Results which, luck aside, confirm the great recovery work carried out by the team despite an R28 that cannot be considered among the season's best cars; that was the fault of a project too similar to the 2007 model, despite an important change of weight distribution with the movement of about two percentage points to the front. The spasmodic research into forward weight movement eased the effect of placing ballast not only in the principal plane of the front wing but also in the end plates (GP of Belgium).
Aerodynamic development was delayed by problems that cropped up during the previous season with work in the wind tunnel and, inevitably, with the tragic loss of the leading man from that sector, Dino Toso, who was replaced by Dick de Beer from BMW.
The other weak points were the notable lower power output of the 8-cylinder Renault engine and greater wear of the rear tyres together with a higher criticality of aerodynamic performance. Due to the reduction in the number of employees at the Viry headquarters, the French engine was behind in both values of maximum power (there was talk of about 35 hp in comparison with the Renault-powered Red Bull and the Toro Rosso with the 8-cylinder Ferrari) and torque.
Renault's season really began in Barcelona with the first major development, almost a B version of the car, with 19 new features. The first concerned, of course, the Red Bull-style "sail", but something new was seen on more or less all sectors of the car. There was a new front wing as in both the bridged flap divided into three quite distinct portions along its length, and its end plates. There were new barge boards behind the front wheels, a new diffuser and brake air intakes. The most impact came from a new development, the addition of a further end plate fixed at about 10 centimetres intervals from the brake intakes, which extended from the limit of the tangent of the wheel onwards, while at the rear area it stopped at the height of the external border of the rim, as required by the regulations.
The GP of Spain was also mechanically an important stage, with the introduction of an inertial damper at the front end and at the rear at Monaco. Another considerable point for Renault in the 2008 season development coincided with the GP of Singapore, in the main due to modifications carried out to the engine.
The numerous aerodynamic changes made during the season did not achieve the same results. The culprit was the front wing, which was considered too pitch-sensitive to height. The developments also concerned the front of the car, which was modified many times, even in its detail, with a view to improving the air flow on the body, such as that directed inside the sidepods and others toward the underbody as far as the diffuser. In China, a new solution was introduced but it was later dropped.

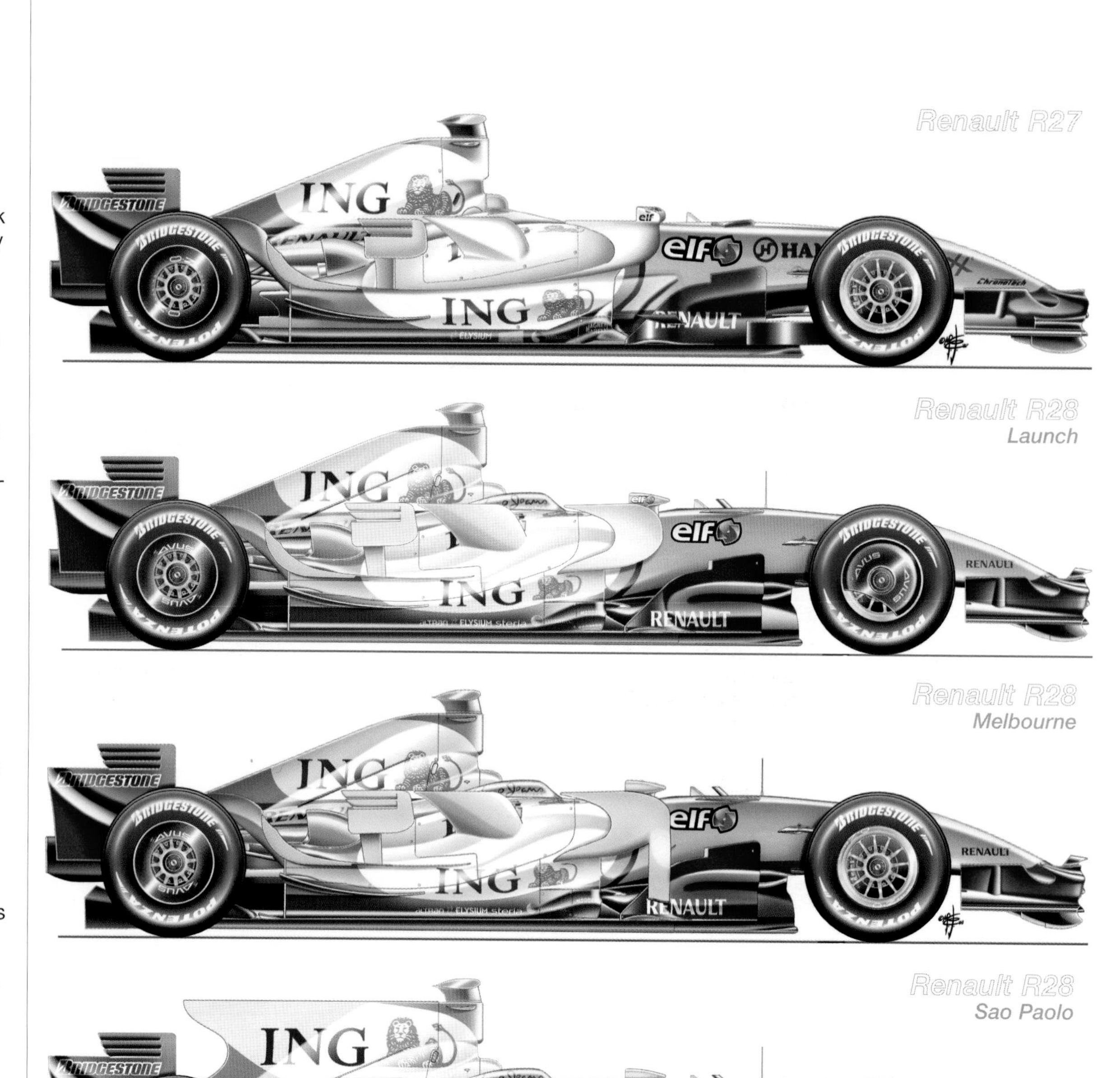

RENAULT

CONSTRUCTORS' CLASSIFICATION	2007	2008	
Position	3°	4°	-1 ▼
Points	51	80	+29 ▲

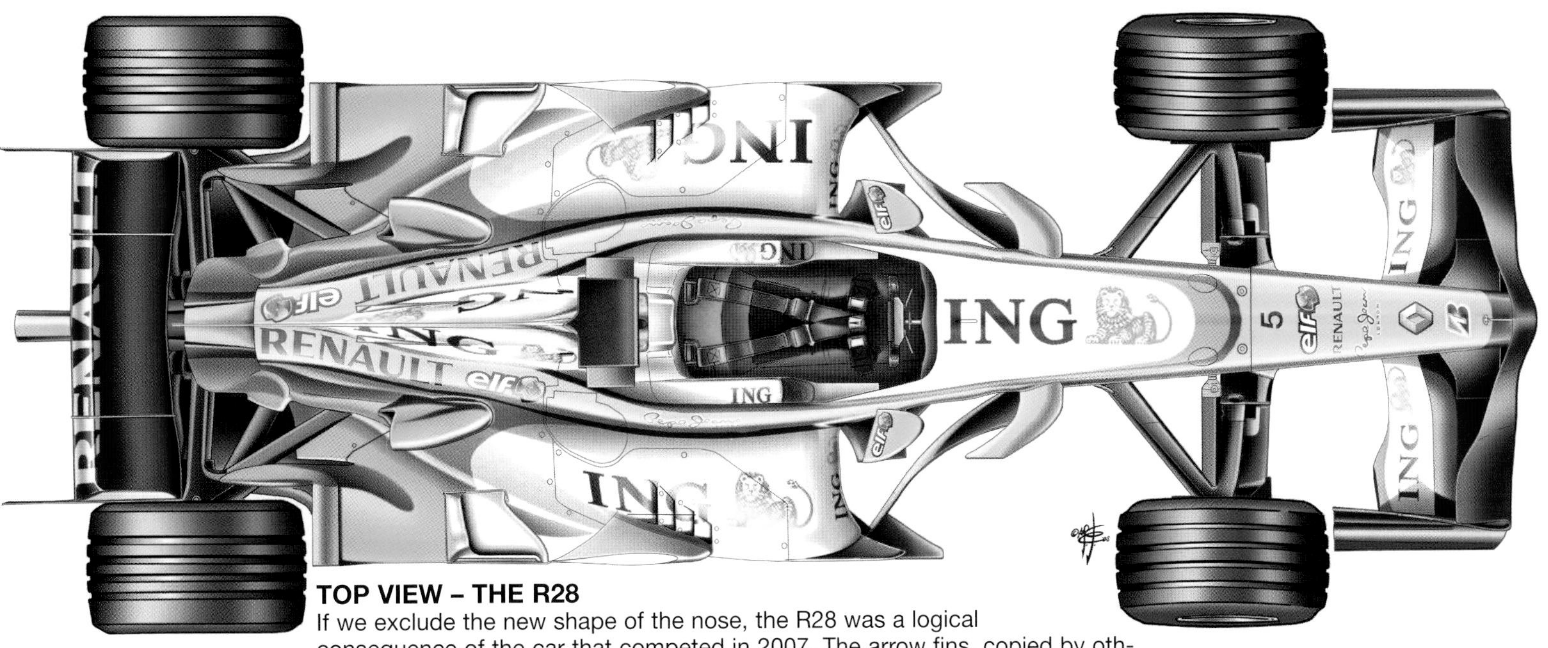

TOP VIEW – THE R28
If we exclude the new shape of the nose, the R28 was a logical consequence of the car that competed in 2007. The arrow fins, copied by others, were retained in the upper part of the chassis as well as the general shape of the body and the large chimneys linked with the lateral flap.

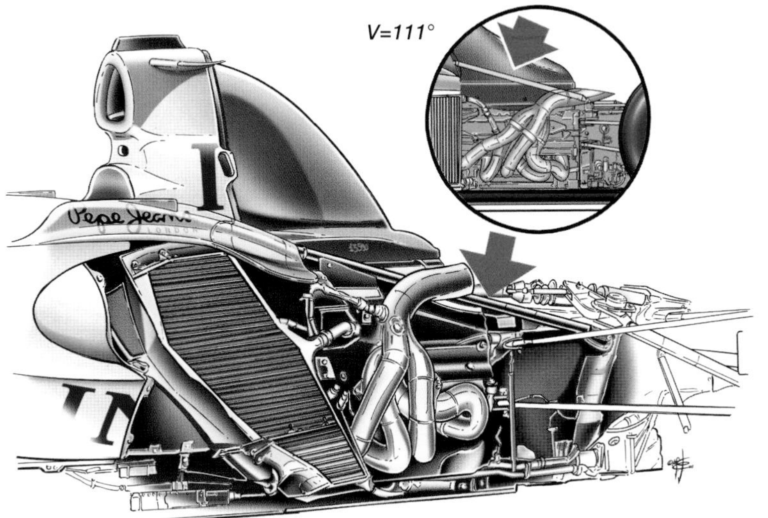

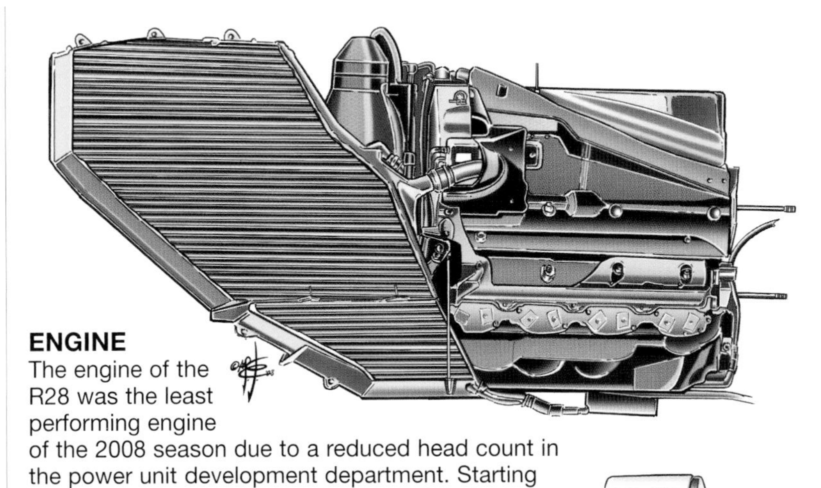

ENGINE
The engine of the R28 was the least performing engine of the 2008 season due to a reduced head count in the power unit development department. Starting from the GP of Singapore, we saw a slight improvement, due in part to the new fuel supplied by Total.

ENGINE FIXING
The Renault was the only car to have a long stiffening arm in carbon fibre, which "linked" the monocoque, engine and gearbox with the task of increasing torsional rigidity.
A solution that was introduced with the 111° power unit and then also transferred to the R24, after which it was always on the French cars.

MONOCOQUE
Stripped, the R28's monocoque taken as a spare and used to simulate refuelling. Note the lateral safety protection that protrudes from the lower part of the chassis. Those in the upper area were hidden inside the initial part of the body. Also in evidence was the monocoque keel with displacement for ballast.

GEARBOX

The transmission group was extremely compact and reasonably developed in height. The suspension push rod acted on the rocker that controlled the torsion bars located in the casting in an almost vertical position, and the anti-roll bar, which in turn operated the third element for the control of the vertical movement. The suspension layout followed that of the R27, with the longitudinal dampers positioned above the gearbox. Note the two wishbone mounts, which were much faired with wing planes.

WING COMPARISON

Below shows how a linked spooned front wing went from that of the R27 to a more concise solution able to ensure greater load, but with a superior criticality. On the R28, this profoundly U-shaped plane was maintained and accentuated, but the classic vertical supports had disappeared in favour of a structure that allowed a better flow of air toward the lower part of the car.

FRONT WING

The most significant new development of the R28 was the front wing mounts with the central part in the nose. In 2007, a heavily spooned principal plane was introduced; square, that of the R28 accentuated the search for maximum downforce, combined with a new way for fixing it to the nose. That was done through an integration of the main plane with the nose. The bridged flap was also retained.

Monaco

Nürburgring

WING FITTING

These two illustrations clearly show how the front wing supports were part of the main plane, with the central cone that was fixed in the plane's central section. That is how they created a good flow of air into the lower zone of the car.

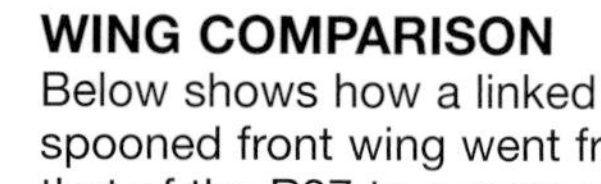

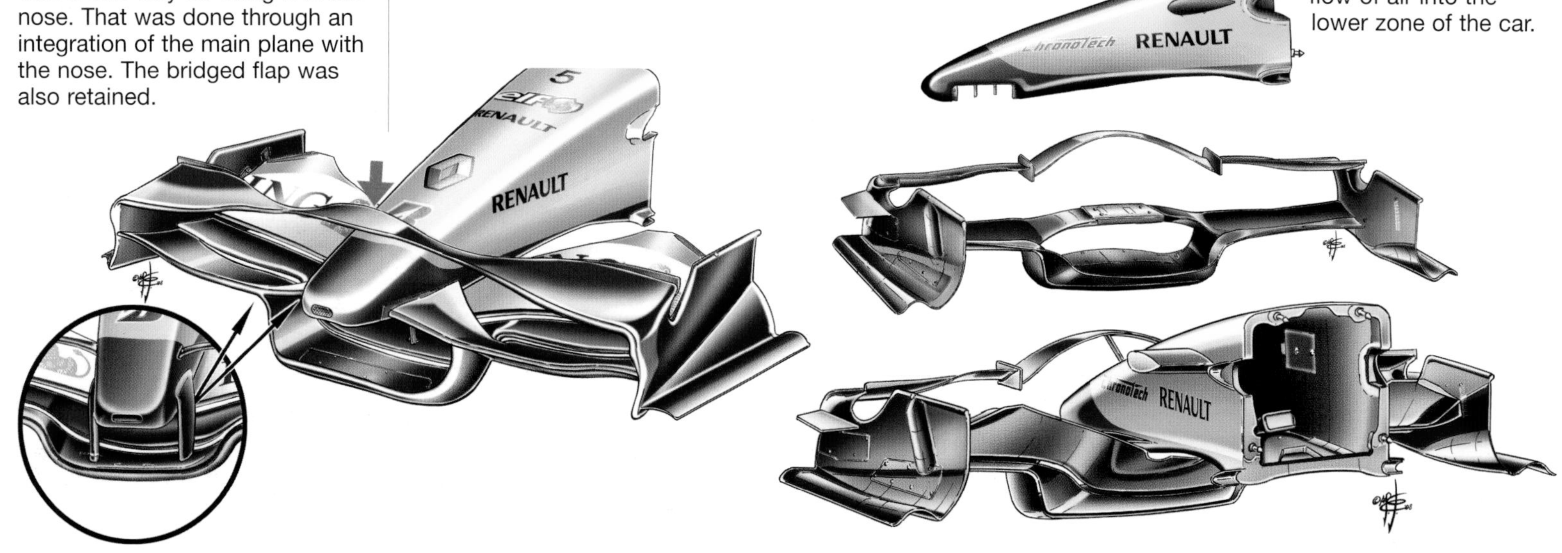

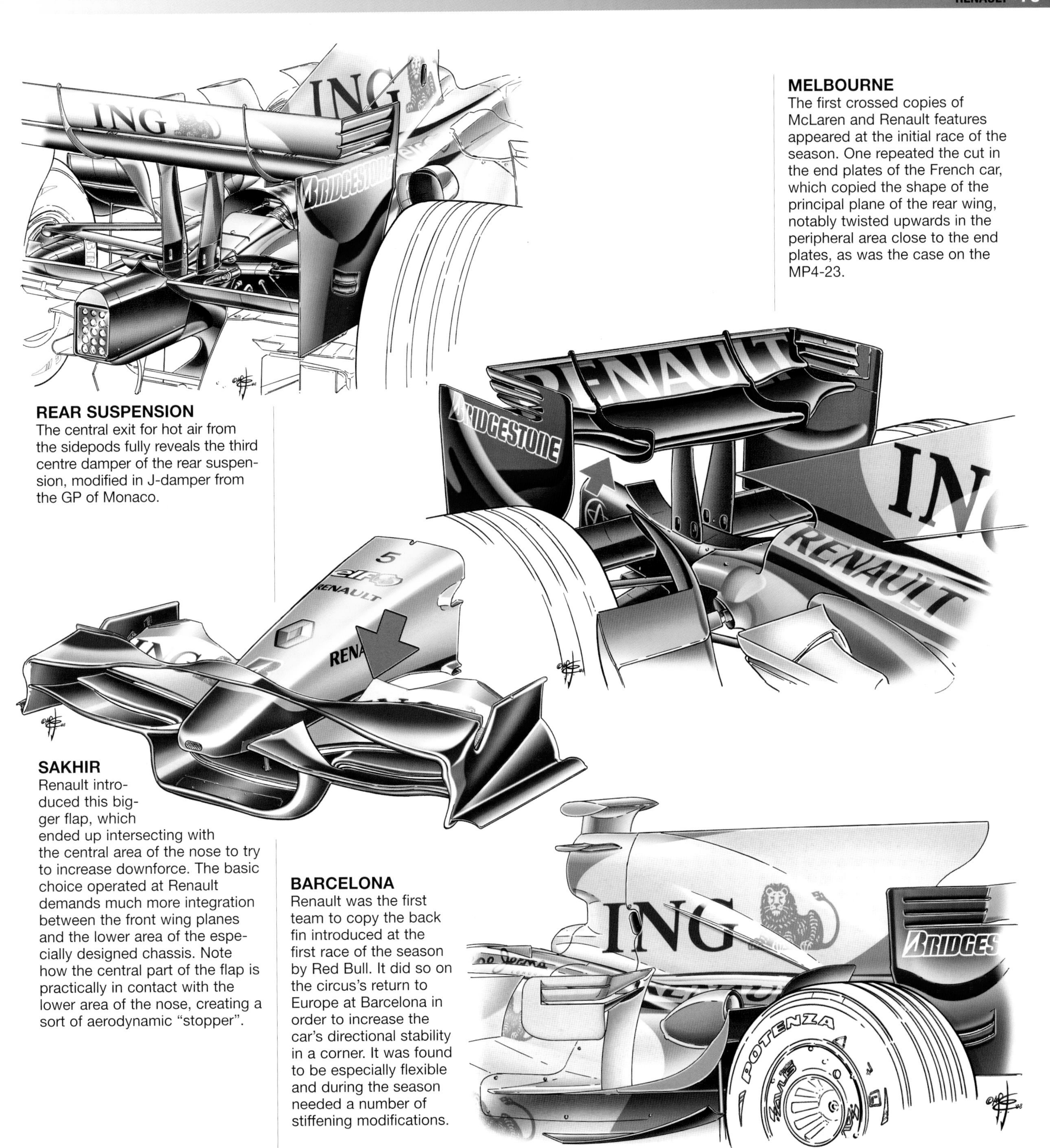

MELBOURNE
The first crossed copies of McLaren and Renault features appeared at the initial race of the season. One repeated the cut in the end plates of the French car, which copied the shape of the principal plane of the rear wing, notably twisted upwards in the peripheral area close to the end plates, as was the case on the MP4-23.

REAR SUSPENSION
The central exit for hot air from the sidepods fully reveals the third centre damper of the rear suspension, modified in J-damper from the GP of Monaco.

SAKHIR
Renault introduced this bigger flap, which ended up intersecting with the central area of the nose to try to increase downforce. The basic choice operated at Renault demands much more integration between the front wing planes and the lower area of the especially designed chassis. Note how the central part of the flap is practically in contact with the lower area of the nose, creating a sort of aerodynamic "stopper".

BARCELONA
Renault was the first team to copy the back fin introduced at the first race of the season by Red Bull. It did so on the circus's return to Europe in order to increase the car's directional stability in a corner. It was found to be especially flexible and during the season needed a number of stiffening modifications.

BARCELONA

Among the R28's many new features introduced at Barcelona was this horizontal fin inside the front wheels. It was a further escalation in the use of aerodynamic devices in the area of the front suspension and is a solution nobody else had applied: it arrives as far forward as far as the tangent of the tyre, while behind it stops at that of the rim, as required by the regulations. The objective was to reduce front wheel turbulence during rotation and align with the air flow directed toward the car's belly and underbody.

MONACO

Two front wings were available to Renault. On the Saturday, the one in the complete illustration was selected, which can be immediately seen as different by the presence of an added external fin with a horizontal progression on the outside: The bulging zone was also different and on this version was concave, while on the one that was not chosen it was convex. The extreme closeness between the central zone of the flap and the lower part of the nose can be clearly seen in the illustration.

ISTANBUL

There were additional variations to the front wing with the adoption of two intermediate end plates in the bridged flap to separate the neutral planes of the central zone (to direct the air flow toward the central part of the car) from the periphery destined to originate download. The central plane was then locked to the nose in deference to a request from FIA, but also to eliminate vibrations which reduced its efficiency.

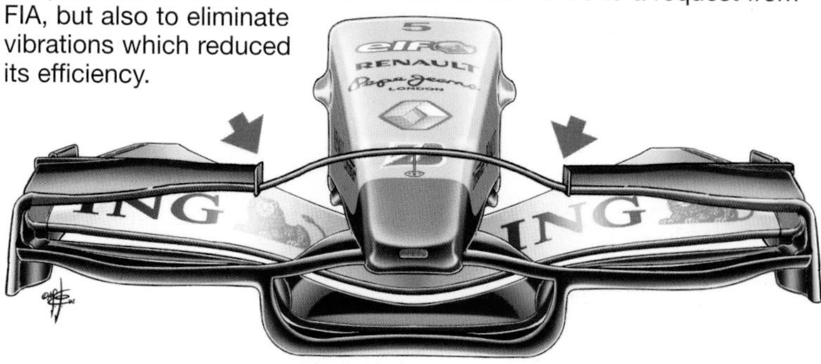

MAGNY COURS

There was a new rear wing for Renault in France. It had its end plates moved about 2-3 centimetres inwards for a different alignment with the front wheels, which reduced the dimensions in the design of the flaps, requiring greater incidence and the addition of a Gurney flap, all with the task of recovering aerodynamic load. A solution against the trends that was dropped by McLaren the previous season. In the 2008 season it was taken up again, revised and corrected by Red Bull, the only team (together with Toro Rosso) with a very wide external section.

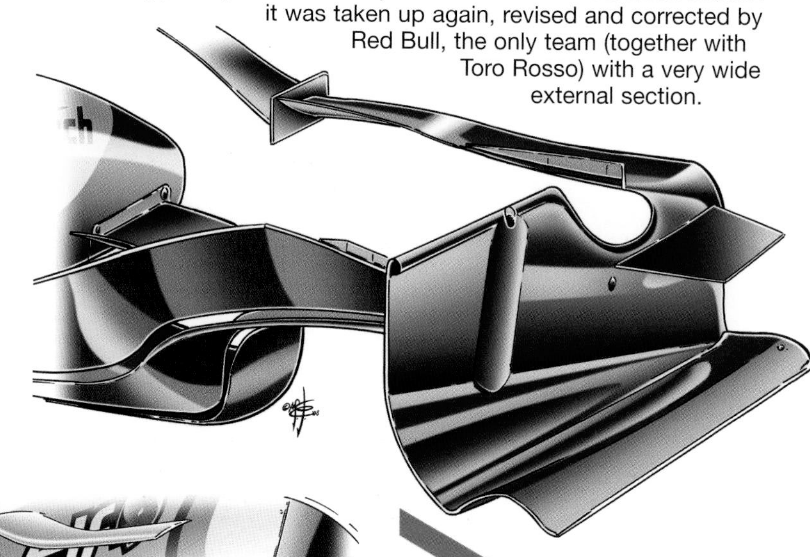

SILVERSTONE

There were new barge boards behind the front wheels for the R28. They were highly concave inwards to improve the extraction of air from the lower part of the car and convey it in the best way possible to the rear diffuser, appropriately directing air inside the car's belly at the same time.

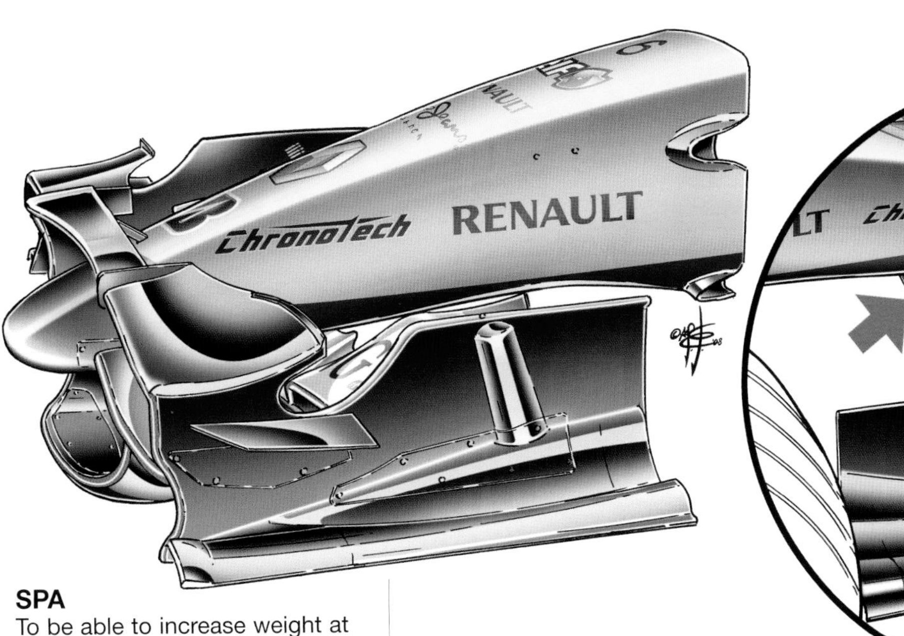

MONZA
Instead of making a new front wing in all its elements, Renault put its money on an adaptation of a traditional one, of which it retained the main plane linked with two flaps with much reduced chord to cut advancement resistance. To balance the rear wing in the best way, this Gurney flap was fitted backwards, slightly reducing front end load.

SPA
To be able to increase weight at the front end, Renault introduced a new placement of ballast in Belgium. It was not only placed in the centre of the principal plane but also inside the end plates for the first time, as show in the diagram.

SINGAPORE
There was a new front wing for the R28, which had a principal plane that was less curved upwards in the area near the end plates. A modification that seemed to have given more stability and uniformity of yield to the wing itself. Note that, from the GP of Belgium, the R28 had a certain quantity of its ballast not only in the central zone of the main plane but also in the end plates.

SHANGAI
Renault was the only team to have taken new developments to China, even if that feature was neither used in qualifying nor the race. It was a new intake that blew fast air from the inner part of the end plates of the rear wheels to the outside to both equilibrate pressure in this zone and reduce the load-bearing effect that originates in this part of the rear wheels; a solution that had been studied in the wind tunnel for some time but reached the circuits at the last moment.

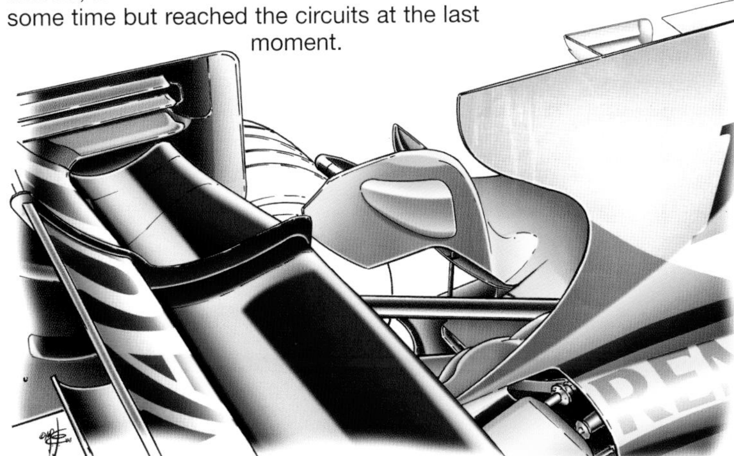

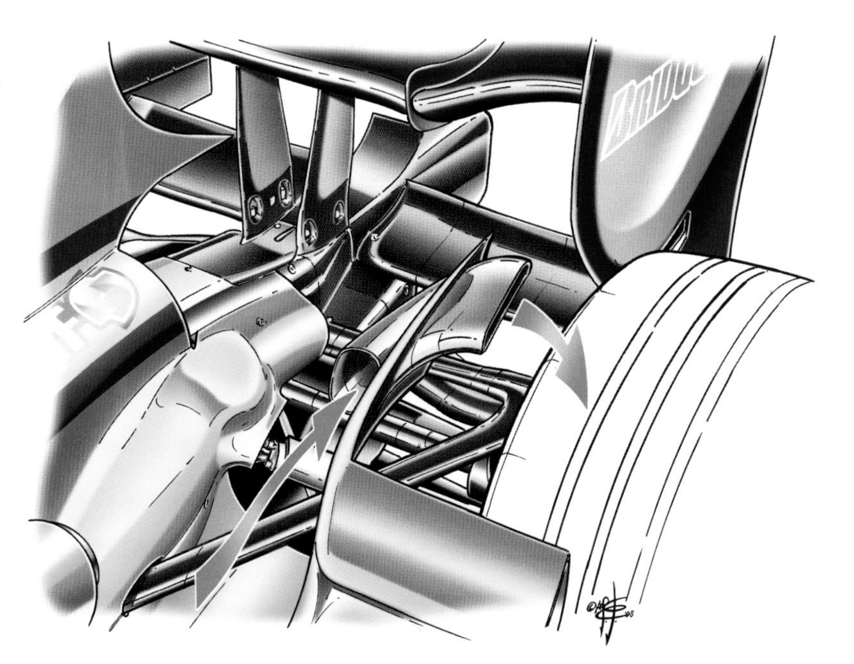

Toyota enjoyed a season of growth, both in terms of Constructors' Championship positions and number of points. This was down to the TF108, a car that for the first time was all-new rather than being a simple evolution of the previous model. Key factors were the aerodynamics and the decision to lengthen the wheelbase by 44 mm which can also be seen as an attempt to distance the front axle and the turbulence generated by the front wheels from the sidepods.
For the first time in Toyota's history the lower section of the sidepods was gently concave and they contained narrower, longer and inclined radiator packs. There was a notable gain both in terms of a reduction in unwanted turbulence and an increase in downforce.
Great attention was paid to the rear section and the exhausts, the work conducted by Dr. Marmorini's group paying off.
The unusual nose was lower and more steeply raked than that of the TF107, while the elegant integration between the bridging flap and the central section of the nose disappeared (curiously, this feature was taken up by Ferrari). The lower part of the wings acted as the mount for the McLaren-type end-plates as well as suspension anchorange points, the suspension layout being completely revised; the elements were less steeply inclined, allowing a more favourable push-rod link angle.

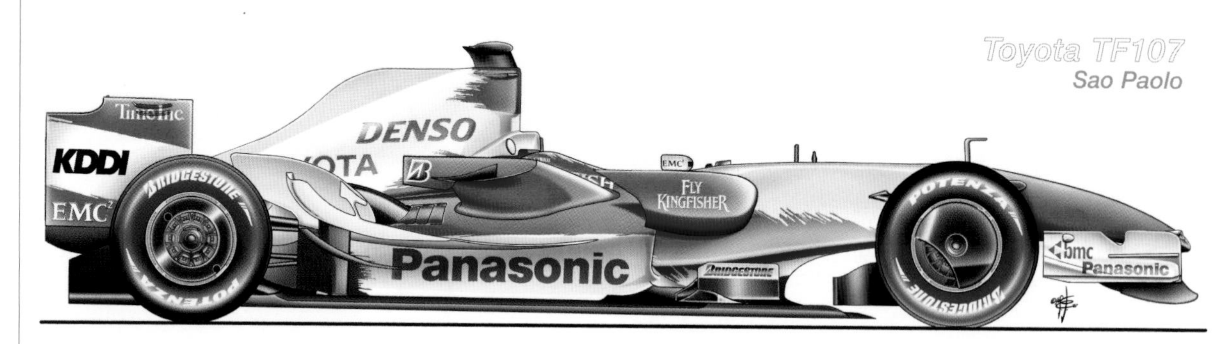

Toyota TF107
Sao Paolo

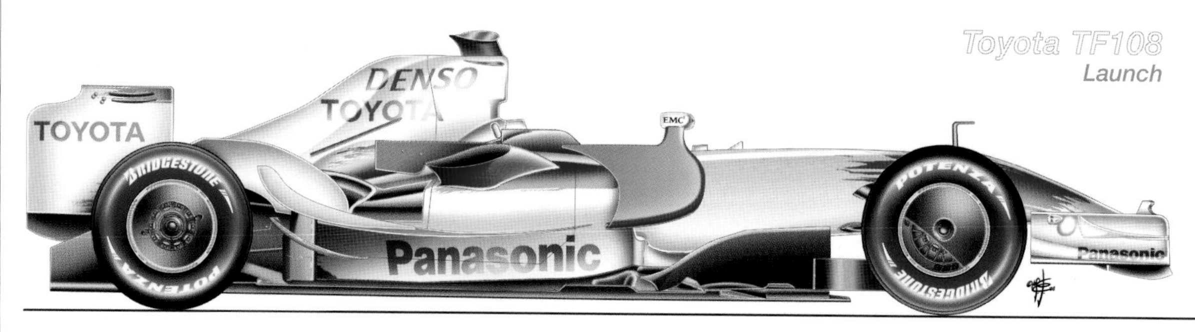

Toyota TF108
Launch

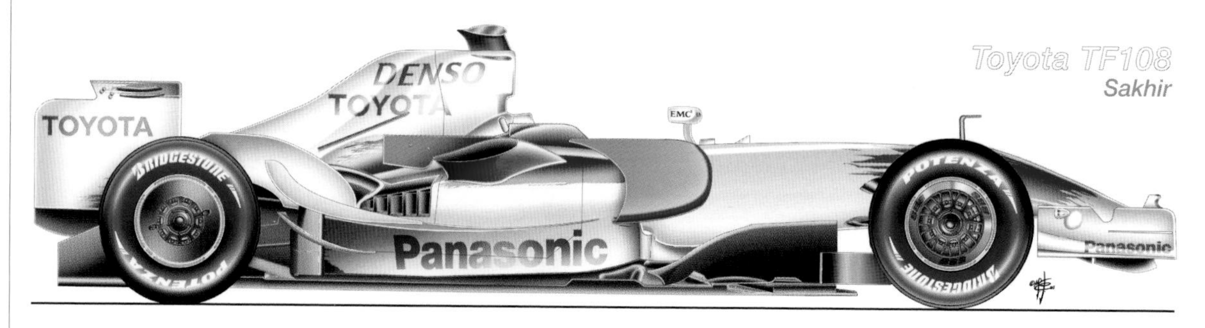

Toyota TF108
Sakhir

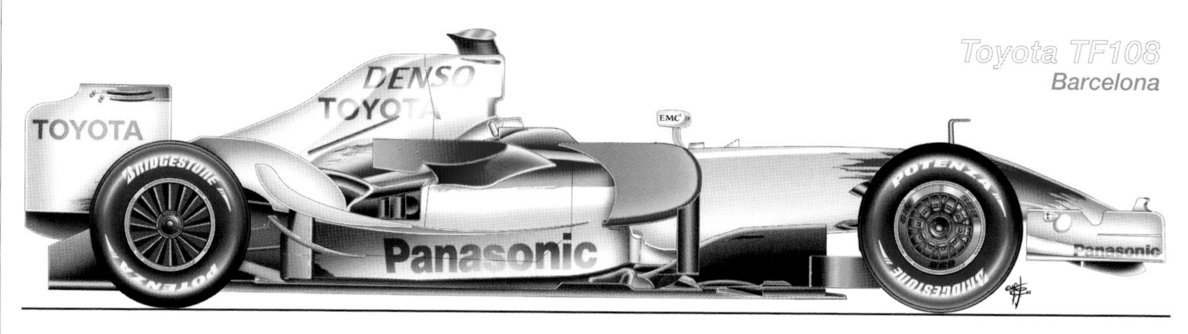

Toyota TF108
Barcelona

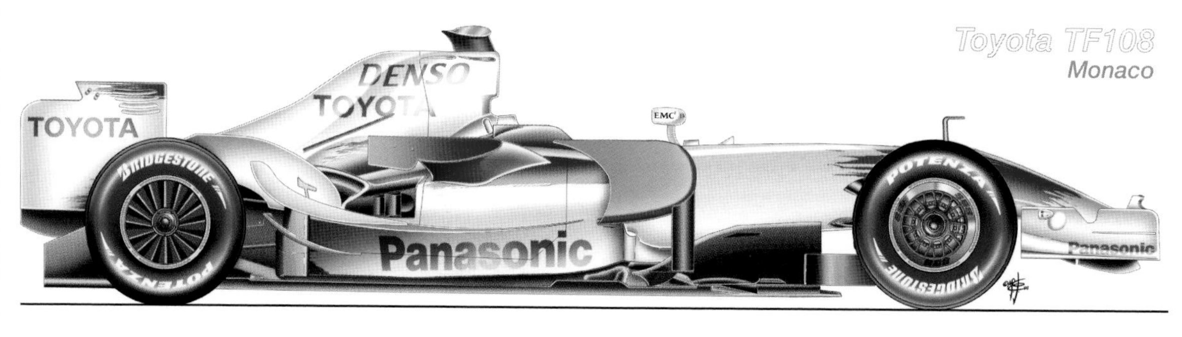

Toyota TF108
Monaco

CONSTRUCTORS' CLASSIFICATION	2007	2008	
Position	6°	5°	+1 ▲
Points	13	56	+43 ▲

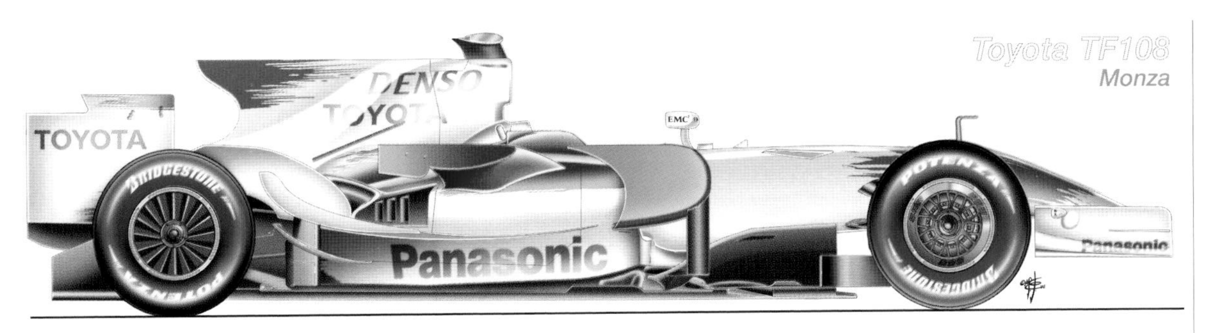

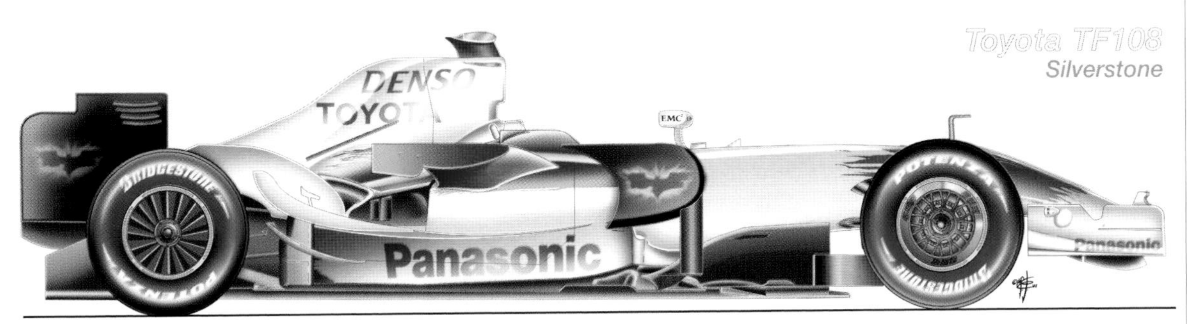

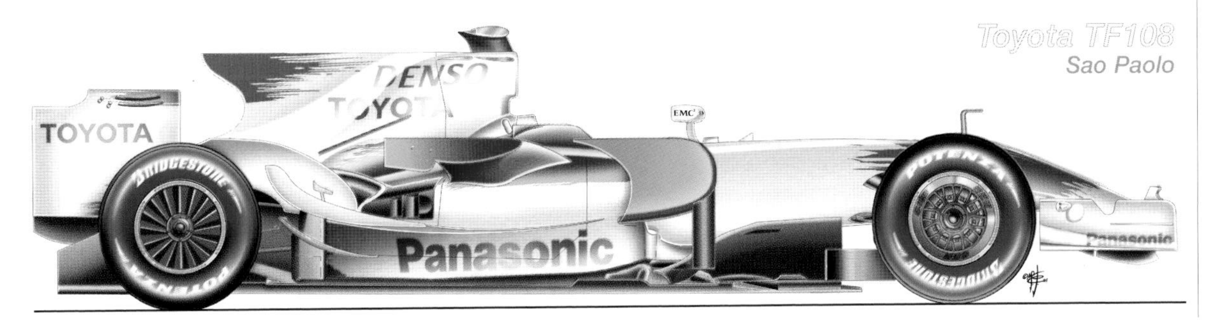

This link was still anchored to the wheel hub but at a point lower than the lower wishbone, in line with current trends.
The gearbox was slightly heavier and longer than that of the TF107. Evolution over the season saw three significant steps: one on the eve of the first race, one in Spain and one in France; subsequently there was continuous development but no substantial modifications.
Following an opening half of the season in which the team led Renault, in the second Toyota was overtaken by the French outfit.
The main problem was that of not being able to warm up the tyres sufficiently except on high-load tracks where the TF108 proved to be more competitive.

SIDEPODS

The TF108 was the first Toyota to feature sidepods with the lower part cut away (1), albeit not to the degree of its leading rivals. The McLaren-style dual end-plates (2) were also new.
The mirrors integrated in the end-plates that had been seen at the car's launch were subsequently eliminated due to visibility problems.

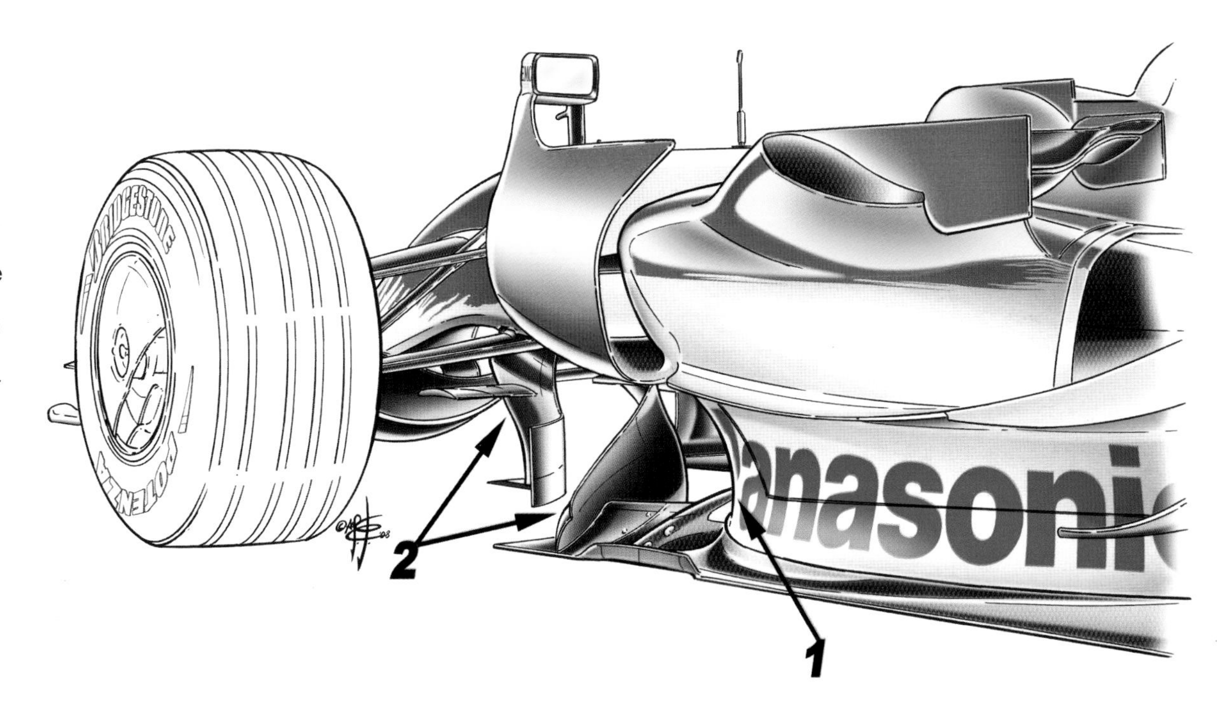

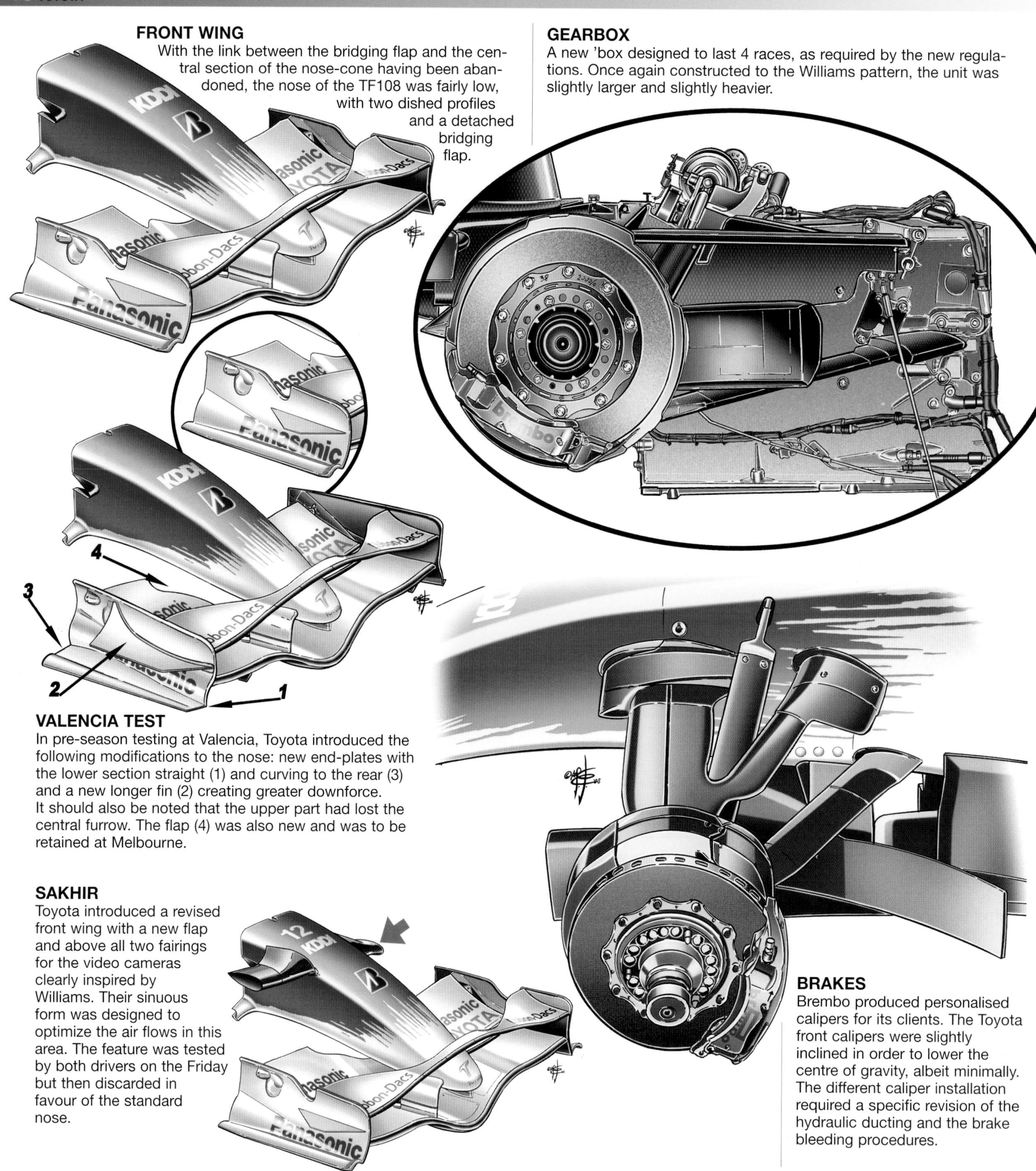

FRONT WING
With the link between the bridging flap and the central section of the nose-cone having been abandoned, the nose of the TF108 was fairly low, with two dished profiles and a detached bridging flap.

GEARBOX
A new 'box designed to last 4 races, as required by the new regulations. Once again constructed to the Williams pattern, the unit was slightly larger and slightly heavier.

VALENCIA TEST
In pre-season testing at Valencia, Toyota introduced the following modifications to the nose: new end-plates with the lower section straight (1) and curving to the rear (3) and a new longer fin (2) creating greater downforce.
It should also be noted that the upper part had lost the central furrow. The flap (4) was also new and was to be retained at Melbourne.

SAKHIR
Toyota introduced a revised front wing with a new flap and above all two fairings for the video cameras clearly inspired by Williams. Their sinuous form was designed to optimize the air flows in this area. The feature was tested by both drivers on the Friday but then discarded in favour of the standard nose.

BRAKES
Brembo produced personalised calipers for its clients. The Toyota front calipers were slightly inclined in order to lower the centre of gravity, albeit minimally. The different caliper installation required a specific revision of the hydraulic ducting and the brake bleeding procedures.

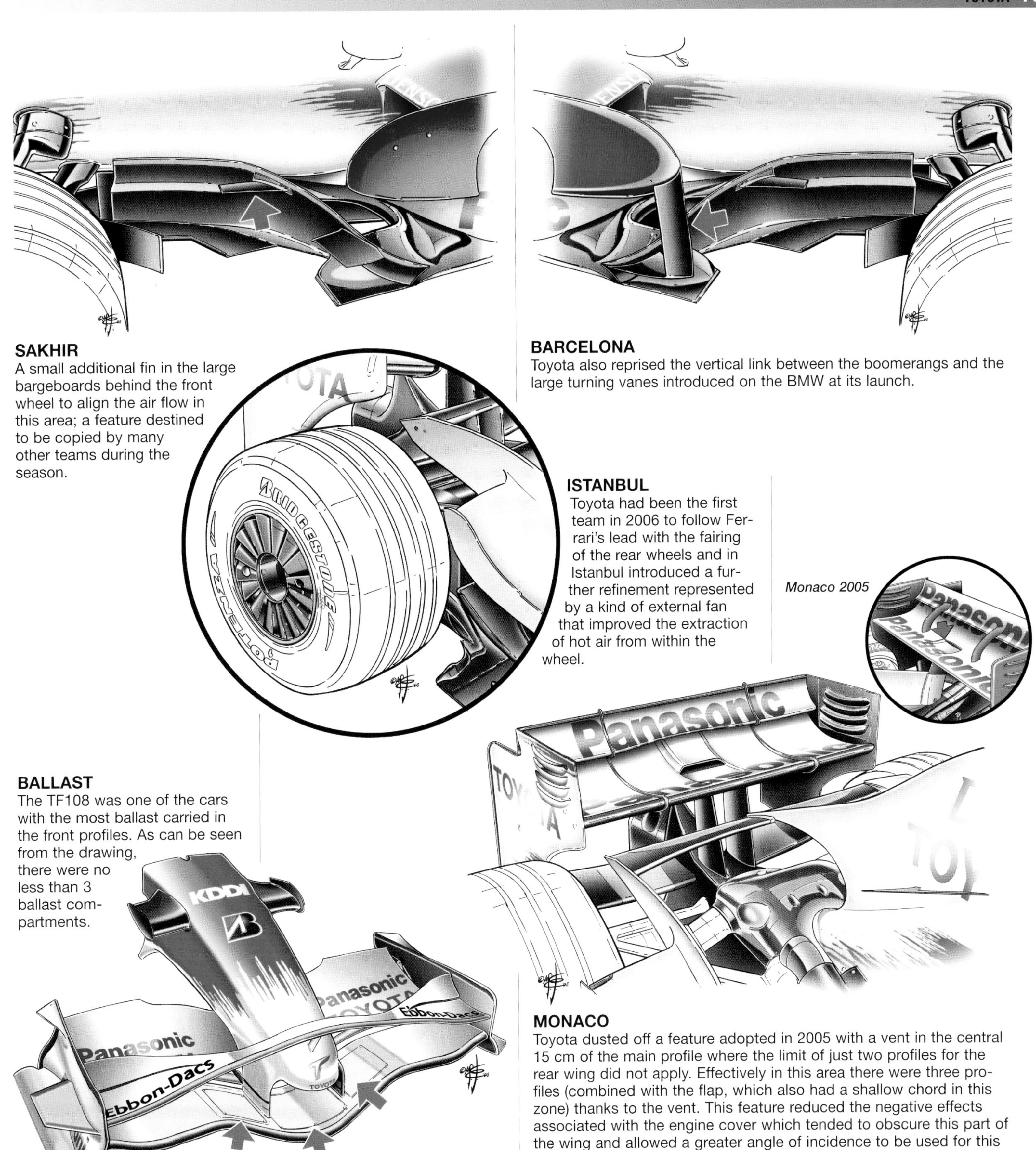

SAKHIR
A small additional fin in the large bargeboards behind the front wheel to align the air flow in this area; a feature destined to be copied by many other teams during the season.

BARCELONA
Toyota also reprised the vertical link between the boomerangs and the large turning vanes introduced on the BMW at its launch.

ISTANBUL
Toyota had been the first team in 2006 to follow Ferrari's lead with the fairing of the rear wheels and in Istanbul introduced a further refinement represented by a kind of external fan that improved the extraction of hot air from within the wheel.

BALLAST
The TF108 was one of the cars with the most ballast carried in the front profiles. As can be seen from the drawing, there were no less than 3 ballast compartments.

MONACO
Toyota dusted off a feature adopted in 2005 with a vent in the central 15 cm of the main profile where the limit of just two profiles for the rear wing did not apply. Effectively in this area there were three profiles (combined with the flap, which also had a shallow chord in this zone) thanks to the vent. This feature reduced the negative effects associated with the engine cover which tended to obscure this part of the wing and allowed a greater angle of incidence to be used for this small section.

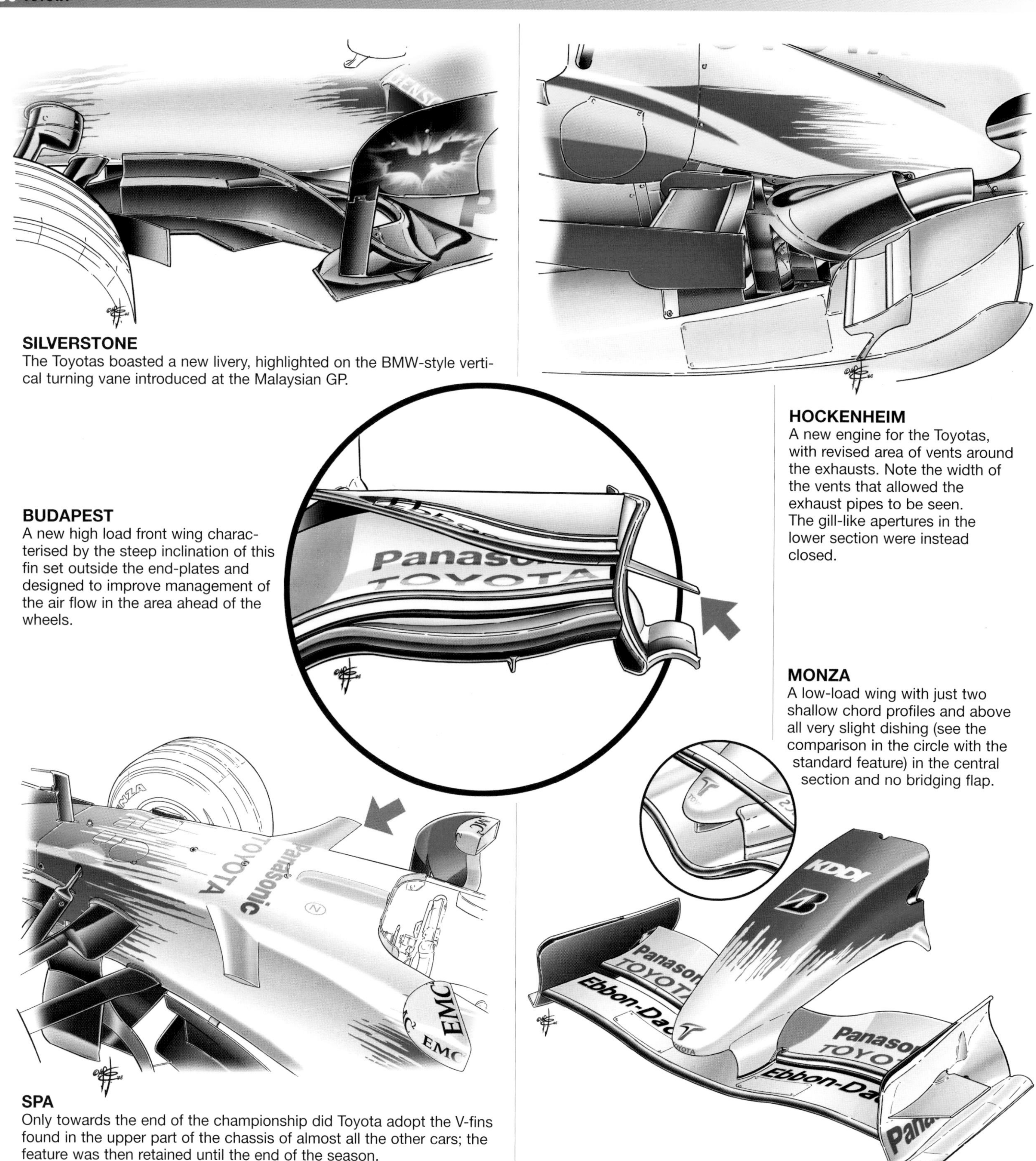

SILVERSTONE
The Toyotas boasted a new livery, highlighted on the BMW-style vertical turning vane introduced at the Malaysian GP.

HOCKENHEIM
A new engine for the Toyotas, with revised area of vents around the exhausts. Note the width of the vents that allowed the exhaust pipes to be seen. The gill-like apertures in the lower section were instead closed.

BUDAPEST
A new high load front wing characterised by the steep inclination of this fin set outside the end-plates and designed to improve management of the air flow in the area ahead of the wheels.

MONZA
A low-load wing with just two shallow chord profiles and above all very slight dishing (see the comparison in the circle with the standard feature) in the central section and no bridging flap.

SPA
Only towards the end of the championship did Toyota adopt the V-fins found in the upper part of the chassis of almost all the other cars; the feature was then retained until the end of the season.

TORO ROSSO

CONSTRUCTORS' CLASSIFICATION	2007	2008	
Position	7°	6°	+1 ▲
Points	8	39	+31 ▲

An exciting season for Toro Rosso, reinforced technically with the grafting onto the Faenza structure of a technician of great experience, Giorgio Ascanelli, ex-track engineer of Gerhard Berger at McLaren and who was recently with Ferrari Although cars that were not notably inferior to the 'headquarters' team Red Bull and with a personnel structure, even if strengthened, of just 173 people – previously it was 158 – Toro Rosso gave itself the satisfaction of finishing before the more aristocratic Red Bull.

And it was precisely in the minor and shrewder development that Ascanelli carried out that made that which, in theory, should have been a handicap become an advantage in a world like Formula 1, where constant development is the basis of success. Putting all the complicated experiments to one side, he concentrated on optimising the handling of the car, especially when the new STR03 arrived in time for the GP of Monaco.

Until then, Toro Rosso had fielded the 2007 car adapted to the new regulations as far as protection at the sides of the driver's head was concerned, as well as new loading for the crash test. The first appearance of the new car at Monaco was certainly not the easiest, both because the city circuit is certainly not the best place to debut a new car, but especially because Bourdais went off in testing at Barcelona, which reduced the spare parts to a bare minimum.

Then the last handicap was the inevitable recession of five places down the grid due to the adoption of the new gearbox, breaking the regulation of the four consecutive races. Despite all of this, a positive result arrived immediately with a 4th by Vettel, who had been driving a car in Red Bull configuration from the start of the season. Obviously, the greatest result was at Monza, where the German's victory took Toro Rosso to the exclusive Olympus of the teams that have won a Formula 1 race. In 2008, there were only five winning teams: as well as the title contenders, Ferrari and McLaren, BMW, Renault and, of course, Toro Rosso won.

Toro Rosso STR02

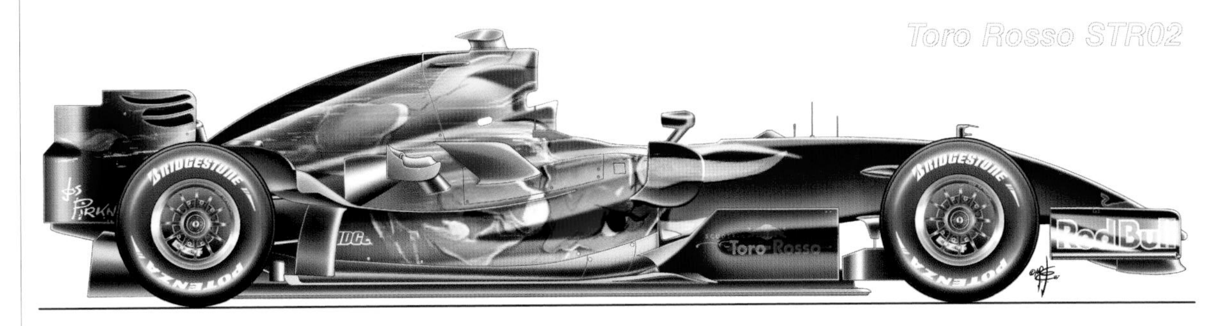

Toro Rosso STR03
Melbourne

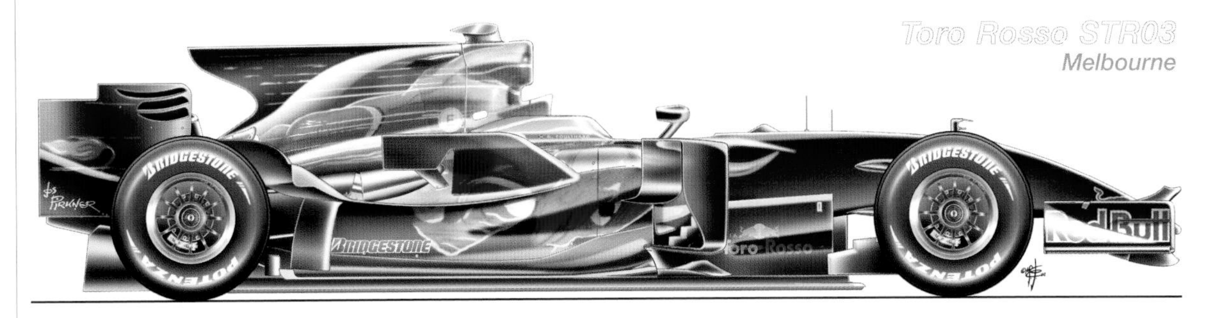

Toro Rosso STR03
Monaco

Toro Rosso STR03
Magny Cours

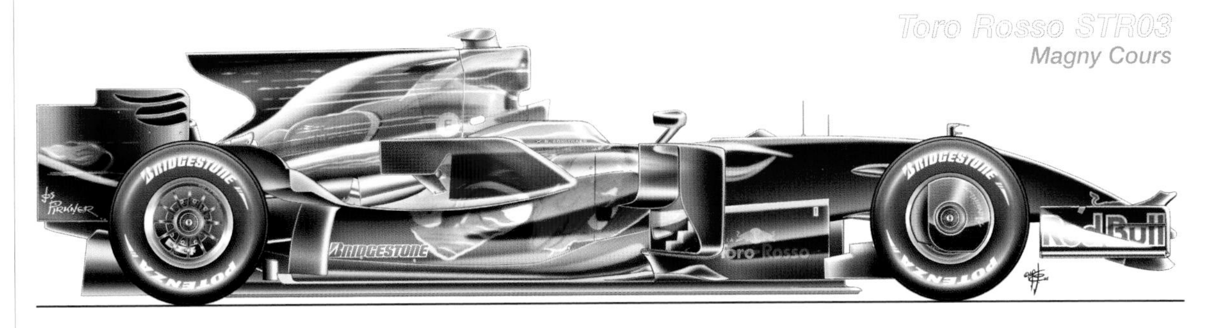

Toro Rosso STR03
Sao Paolo

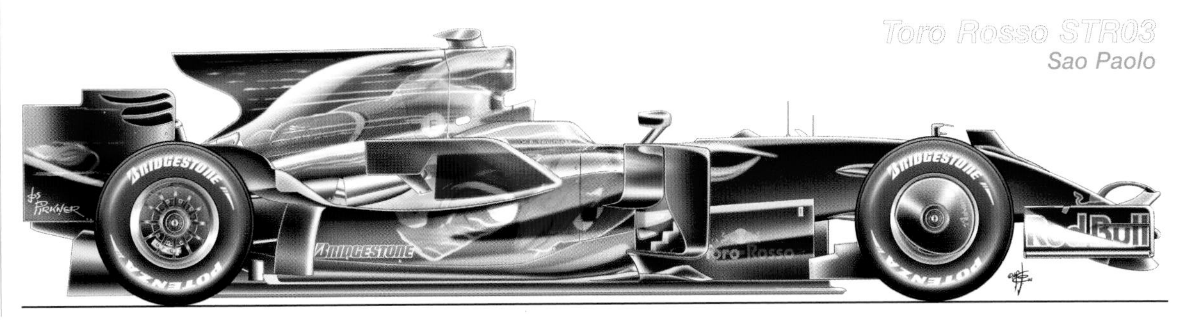

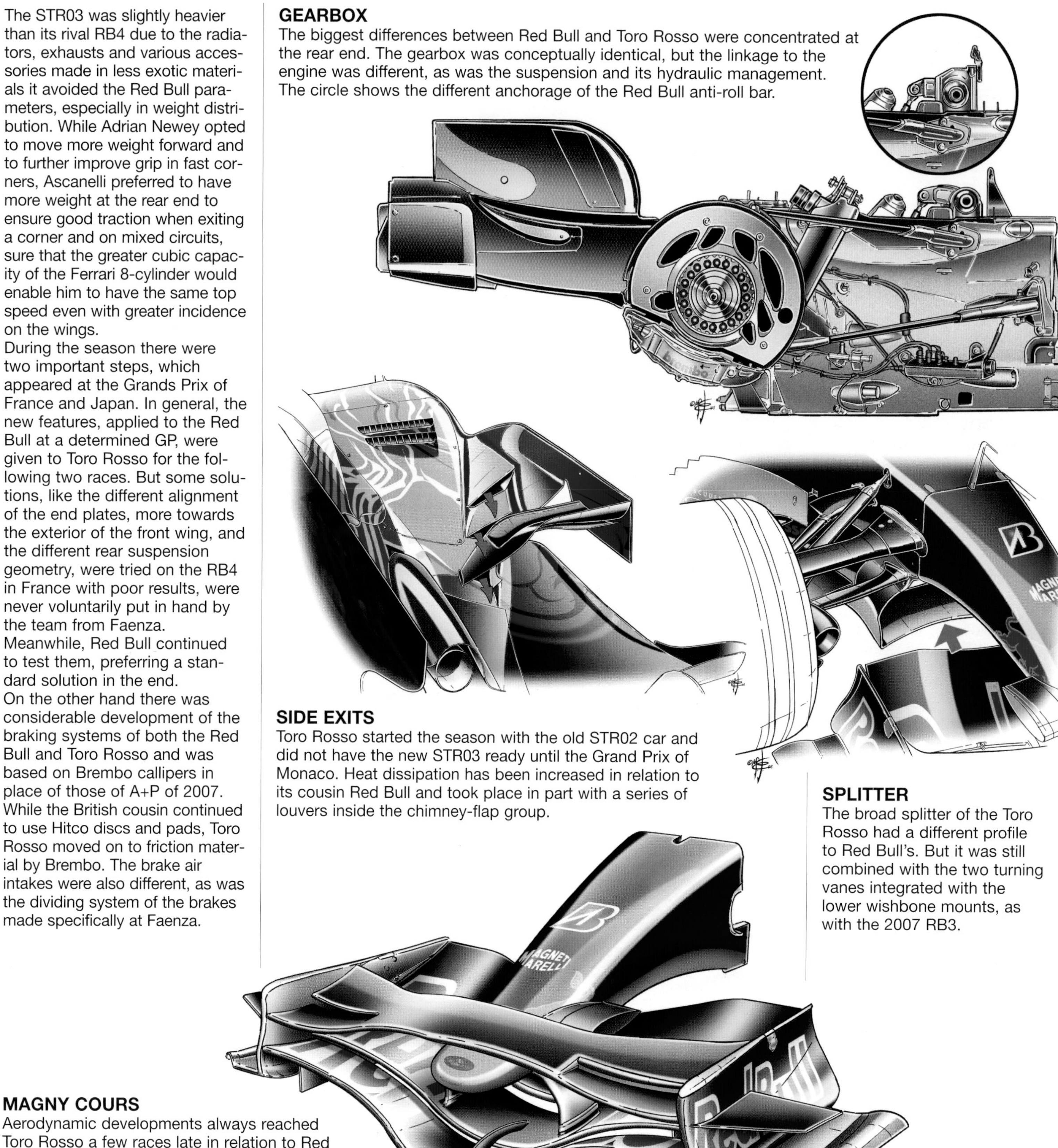

The STR03 was slightly heavier than its rival RB4 due to the radiators, exhausts and various accessories made in less exotic materials it avoided the Red Bull parameters, especially in weight distribution. While Adrian Newey opted to move more weight forward and to further improve grip in fast corners, Ascanelli preferred to have more weight at the rear end to ensure good traction when exiting a corner and on mixed circuits, sure that the greater cubic capacity of the Ferrari 8-cylinder would enable him to have the same top speed even with greater incidence on the wings.
During the season there were two important steps, which appeared at the Grands Prix of France and Japan. In general, the new features, applied to the Red Bull at a determined GP, were given to Toro Rosso for the following two races. But some solutions, like the different alignment of the end plates, more towards the exterior of the front wing, and the different rear suspension geometry, were tried on the RB4 in France with poor results, were never voluntarily put in hand by the team from Faenza. Meanwhile, Red Bull continued to test them, preferring a standard solution in the end.
On the other hand there was considerable development of the braking systems of both the Red Bull and Toro Rosso and was based on Brembo callipers in place of those of A+P of 2007. While the British cousin continued to use Hitco discs and pads, Toro Rosso moved on to friction material by Brembo. The brake air intakes were also different, as was the dividing system of the brakes made specifically at Faenza.

GEARBOX
The biggest differences between Red Bull and Toro Rosso were concentrated at the rear end. The gearbox was conceptually identical, but the linkage to the engine was different, as was the suspension and its hydraulic management. The circle shows the different anchorage of the Red Bull anti-roll bar.

SIDE EXITS
Toro Rosso started the season with the old STR02 car and did not have the new STR03 ready until the Grand Prix of Monaco. Heat dissipation has been increased in relation to its cousin Red Bull and took place in part with a series of louvers inside the chimney-flap group.

SPLITTER
The broad splitter of the Toro Rosso had a different profile to Red Bull's. But it was still combined with the two turning vanes integrated with the lower wishbone mounts, as with the 2007 RB3.

MAGNY COURS
Aerodynamic developments always reached Toro Rosso a few races late in relation to Red Bull, like this new front wing with an axle McLaren-type flap blown at the sides and fitted for the GP of France. It was seen on the Red Bull at Monaco.

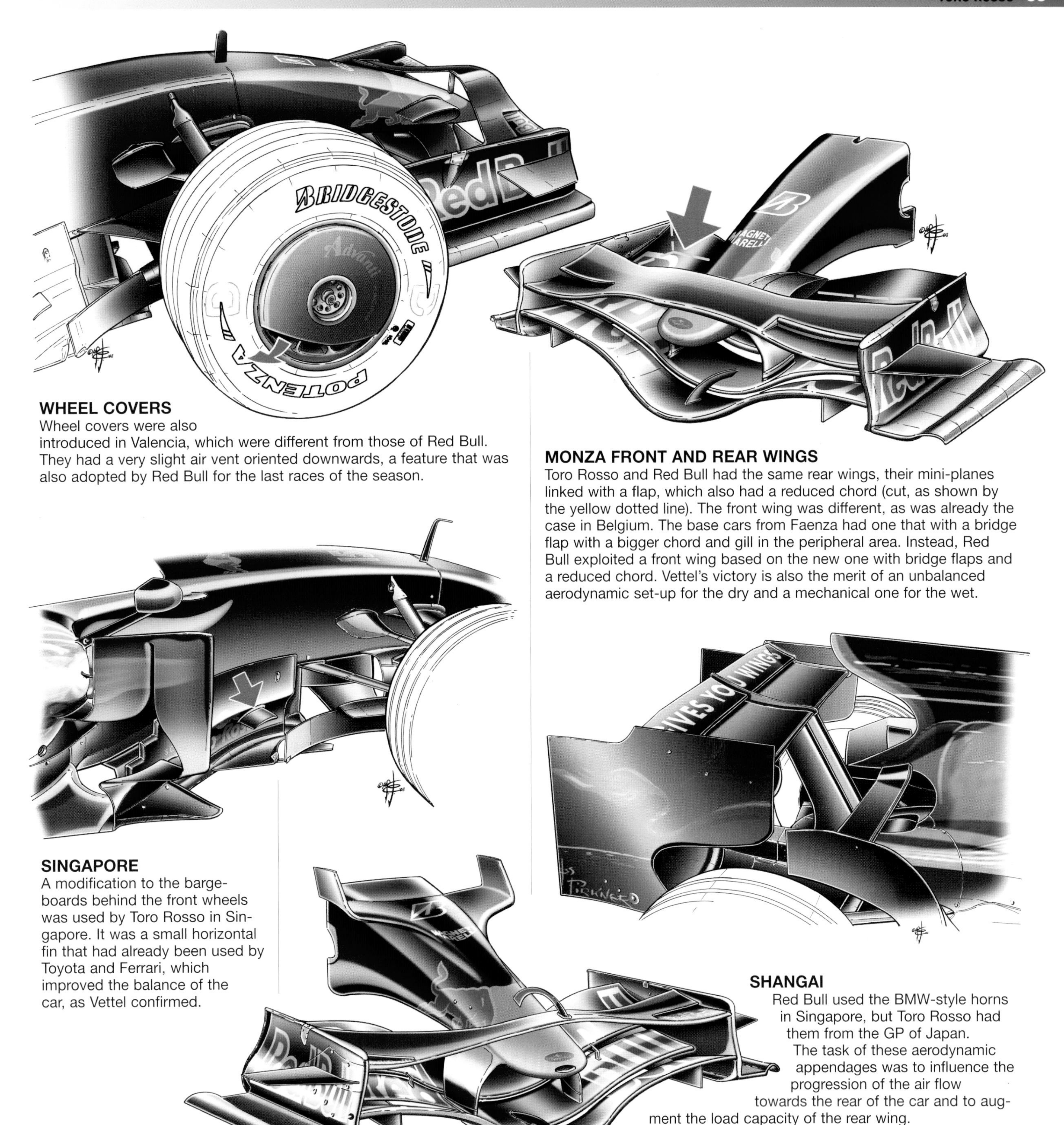

WHEEL COVERS

Wheel covers were also introduced in Valencia, which were different from those of Red Bull. They had a very slight air vent oriented downwards, a feature that was also adopted by Red Bull for the last races of the season.

MONZA FRONT AND REAR WINGS

Toro Rosso and Red Bull had the same rear wings, their mini-planes linked with a flap, which also had a reduced chord (cut, as shown by the yellow dotted line). The front wing was different, as was already the case in Belgium. The base cars from Faenza had one that with a bridge flap with a bigger chord and gill in the peripheral area. Instead, Red Bull exploited a front wing based on the new one with bridge flaps and a reduced chord. Vettel's victory is also the merit of an unbalanced aerodynamic set-up for the dry and a mechanical one for the wet.

SINGAPORE

A modification to the barge-boards behind the front wheels was used by Toro Rosso in Singapore. It was a small horizontal fin that had already been used by Toyota and Ferrari, which improved the balance of the car, as Vettel confirmed.

SHANGAI

Red Bull used the BMW-style horns in Singapore, but Toro Rosso had them from the GP of Japan. The task of these aerodynamic appendages was to influence the progression of the air flow towards the rear of the car and to augment the load capacity of the rear wing.

The 2008 season did not produce the results Red Bull had hoped for from the RB4, Adrian Newey's second car for the Anglo-Austrian team. But to try and resolve the problems of the previous RB3 in the best possible way, Geoff Willis, ex-technical director of Honda who cut his teeth in the Patrick Head school and a man of considerable experience, was engaged midway through 2007.
The result was two positions lower down in the world championship table, despite a few steps forward in terms of points, but above all the humiliation of having been overtaken by its poor relation Toro Rosso, which was even able to win a race with Sebastian Vettel at Monza.
The fault, in part, was the decision of Adrian Newey's to drop the 8-cylinder Ferrari engine, which required bigger radiator packs, in favour of the 8-cylinder Renault, which in the end enabled the team to earn a slight improvement in terms of bulk of installation and weight, but it generated about 35 hp less compared to the Maranello power unit. The biggest difference between the two cars was almost all in the rear end and with the engine installations, like the oil reservoir, the hydraulic system of the gearbox and engine and the installation of the radiators.
The RB4 was not a revolutionary car, but a logical development of the RB3 with tree objectives that had to be achieved: better reliability, the ability to be able to move more ballast up front and more efficient aerodynamics.
Newey followed the trend of his McLarens with a fairly long and narrow nose, inclined slightly downwards and with multiplanes at the front wing linked with bargeboards moved towards the centre with the task of creating a more favourable alignment in large, fast corners.
That was where the RB4 did, in fact, show it was much at ease and had an advantage over rival Toro Rosso, in part due to a weight distribution further ahead.
The car was, however, in difficulty in mixed and slow sections of the circuits. The big fin on the engine cover was new for today's Formula 1, even if it took up again a technique seen on the McLaren MP4/10 of 1995 and even earlier on the 1971 Ferrari B2; a feature that set a fashion for others and about which we shall speak in the aerodynamics chapter.
Development was notable until mid-season, even if sometimes certain features, especially concerning the rear suspension, were rejected by the drivers. The car's Toro Rosso cousin, with less development but a more careful set-up, therefore,

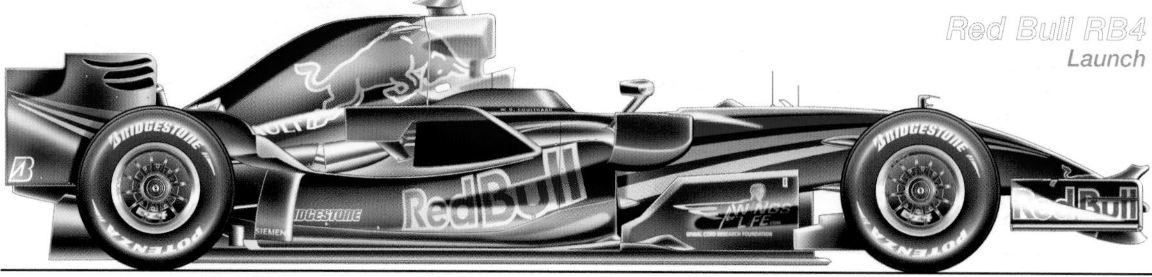

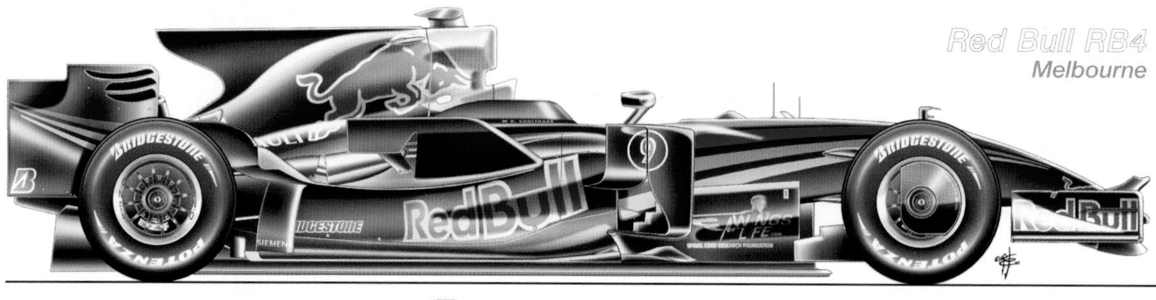

RED BULL

CONSTRUCTORS' CLASSIFICATION	2007	2008	
Position	5°	7°	-2 ▼
Points	24	29	+5 ▲

showed it had the advantage. Newey immediately took on board the vertical connection between the turning vanes and the boomerangs in the area above the sidepods, introduced by BMW, as he did the horns in the upper part of the nose. Coulthard's accident in Malesia created some controversy with FIA ready to request complete documentation from the team before allowing it to race again. So Red Bull provided the Federation's verifiers with a dossier in which it was established that the break point took place around a load of 3.1 tonnes. It should be pointed out that every single piece of the car had precise specifications concerning dimensioning, the load it must exceed and its vital cycle.
The steering link that had become unstuck (perhaps because of the heat emitted inside the brake drum duct) must resist a tension and compression of at least 1,000 kg.
This value, depending on the team and the vital cycle (about 3,000 km or four races for the suspension arms) is almost doubled for safety reasons.
Before moving on to the productive stage, every piece is subjected to a very severe verification test and compared with the values of the previous component.
Returning to the RB4, in the second half of the season development was slowed, even if new features were introduced at the GP of Singapore. Newey had, in fact, opted to concentrate 100% on the design of the 2009 car and, in general, on the changes of regulation, as a result of which the British genius has always been able to come up with something new.
In the comparison between the profiles, we decided to use the special livery designed for Brazil and David Coulthard's last race, retiring after his 246th GP after 14 years in Formula 1.

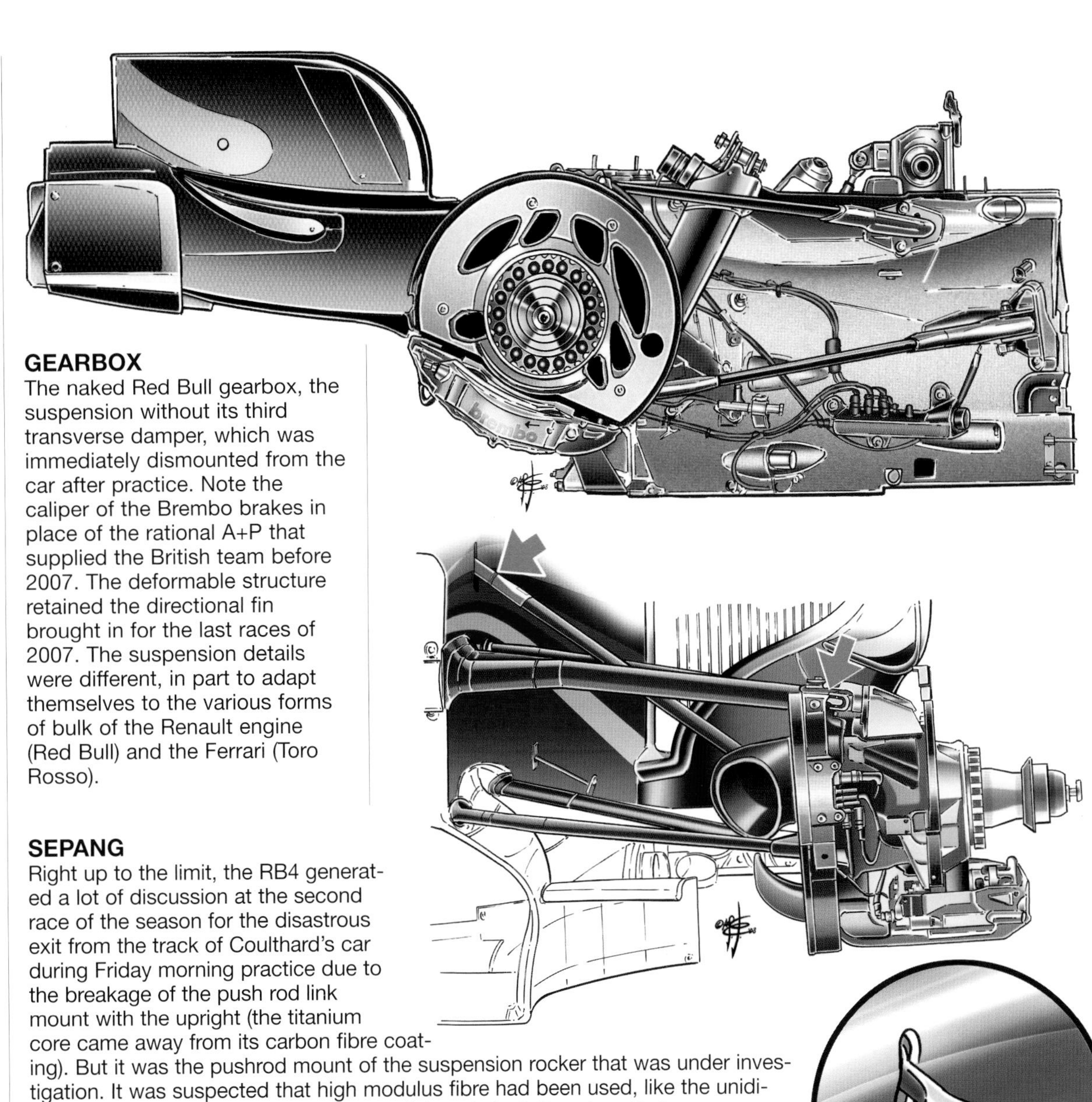

2007

2008

GEARBOX

The naked Red Bull gearbox, the suspension without its third transverse damper, which was immediately dismounted from the car after practice. Note the caliper of the Brembo brakes in place of the rational A+P that supplied the British team before 2007. The deformable structure retained the directional fin brought in for the last races of 2007. The suspension details were different, in part to adapt themselves to the various forms of bulk of the Renault engine (Red Bull) and the Ferrari (Toro Rosso).

SEPANG

Right up to the limit, the RB4 generated a lot of discussion at the second race of the season for the disastrous exit from the track of Coulthard's car during Friday morning practice due to the breakage of the push rod link mount with the upright (the titanium core came away from its carbon fibre coating). But it was the pushrod mount of the suspension rocker that was under investigation. It was suspected that high modulus fibre had been used, like the unidirectional M46J and the M55J with a modulus of elasticity values that were very high; they guarantee extreme rigidity, but are at risk under dynamic stress with particular angles of impact. And in effect, Red Bull had passed in a contrary sense on the internal step of the kerb, therefore with abnormal angle and direction. Obviously, compared to others like the T 800 and the T 400, these fibres provide a notable rigidity together with a good reduction in weight: two factors that have their positive effect on the efficiency of the suspension, all to the advantage of mechanical grip. In the end, the accident was filed away as having originated from a design problem and not from a single defective component. The push rod mount of the Red Bull, which was different from all the others and from that of last year (visible in the circle on the right), was regulated with spacers in the narrow area. For the Saturday, the team replaced this component with another in steel of modified shape.

BARCELONA

The first aerodynamic development package was introduced at Barcelona, in which the connection between the boomerangs and turning vanes in the area in front of the sidepods was the most evident aspect. In practice, Red Bull and Toyota copied the feature brought in by BMW and which had already been seen from the first race of the year on the Renault. The new fin on the RB4 also required a modification to the barge boards, which became higher.

MONACO

Two wings available to the Red Bulls, but the one selected was new and had a bridge that had a bigger chord and two McLaren-style vents in the peripheral area to increase the negative lift.

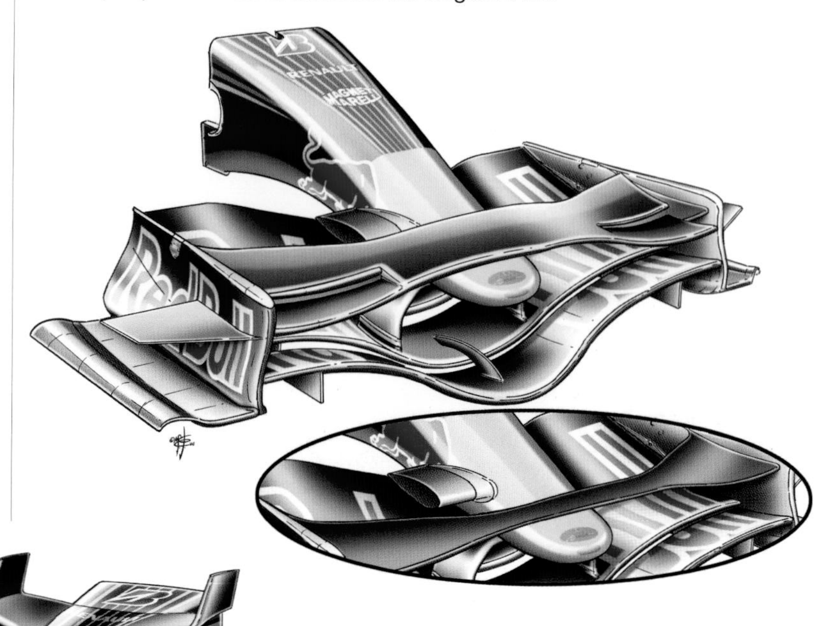

SINGAPORE

Red Bull brought in its last development of the season at Singapore. It was practically a photocopy of the BMW horns applied to the sides of the nose to redirect the air flow toward the central part of the car, a feature later adopted by its relation Toro Rosso in the next race.

SHANGAI

For the entire season, both Red Bull and Toro Rosso kept a more internal alignment of the end plates, even if Adrian Newey's team experimented mid-season with a more standard configuration. In China, the squad had two different types of bargeboards available to it, which were different from each other in the semi-horizontal part. The version in the illustration was the one selected, as it gave a better seal to the ground and, therefore, greater efficiency of the wing itself.

SILVERSTONE

Red Bull once again tried the new wing introduced in France, modified with an added second separator fin in the lower zone, but preferred the traditional solution (below) in the end with wide end plates, which permitted a better alignment with the tyres when steering in the long corners at Silverstone, ensuring greater stability.

ENGINE FAIRING

The maniacal care taken with aerodynamic research can be seen in this mini-wing plane obtained in the thermal screening that is wrapped around the engine. It acts as a hot air extractor, also separating the flows that come from the radiators and the heat of the engine.

WILLIAMS

CONSTRUCTORS' CLASSIFICATION	2007	2008	
Position	4°	8°	-4 ▼
Points	33	26	-7 ▼

This was a bleak season for Williams as the team dropped no less than 4 places in the Constructors' standings, despite having a car that had appeared to be competitive, especially after its debut at Melbourne.
In spite of the strong structural links with the British team's 2007 car, many features of which it retained, the new FW30 was clearly influenced by the two pack leaders Ferrari and McLaren in certain areas.
The similarity with the 2007 MP4-22 lay in the integration of the vanes in the upper part of the side-pods with the chimneys.
The family feeling is generated by the shape of the fairly low nose, combined with three dished profiles and bridging flaps (again following McLaren's lead) and by the high, rounded sidepods.
Still at the front end, mention should be made of the integration of the video camera with sophisticated fairings designed to align the flow of air from the front wing. The wheelbase was slightly longer (+39 mm).
Great attention was paid to the rear end, with features borrowed from the Ferrari like the central section of the diffuser and the position of the hydraulic system radiator, cooled via an internal duct from the engine air intake, as on the F2007. The Ferrari also inspired the movement of the lateral deformable structures from the chassis to the stepped underbody and the shifting outboard of the front suspension upper wishbone mount.
The fluid dynamics within the sidepods, combined with the work conducted in the wind tunnel and the full use of inertia dampers made the FW30 7/10 quicker than the 2007 car in the first track tests, but this was not enough when faced with the competition. More at its ease on slower tracks, the FW30 struggled through fast corners, as demonstrated in the races at Silverstone and Spa.

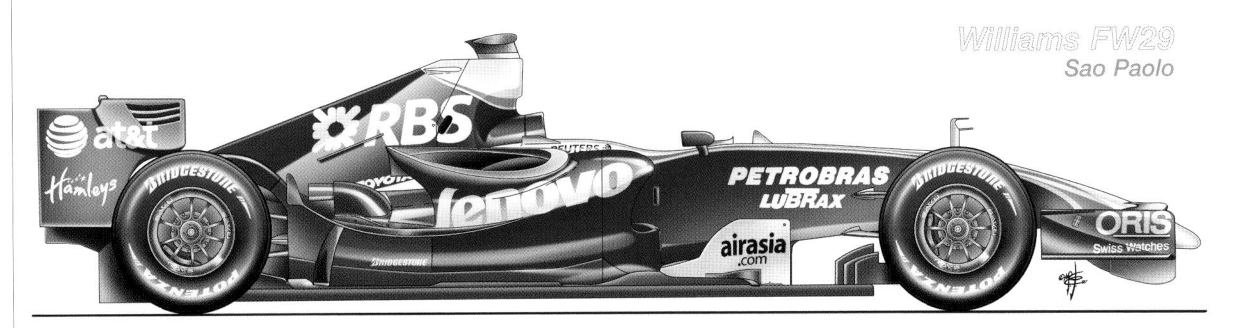

Williams FW29
Sao Paolo

Williams FW30
Launch

Williams FW30
Barcelona

Williams FW30
Montreal

Williams FW30
Magny Cours

On the other hand, Melbourne, Monaco, Montreal and Singapore proved to be favourable tracks. The adoption of a more aggressive extractor profile, with the integration of the lower suspension wishbone with the flap function in order to increase downforce, meant that the team was obliged to produce front wings capable of balancing the front and rear ends. This led to intensive development in the first part of the season producing a notable variety of front wings, but at a certain point this development was put on hold in favour of working on the 2009 car. Even at the start of the 2008 season, research had been divided equally between the FW30 and the future FW31; gradually the new car began to take precedence as the team attempted to ensure that Williams would be in a position to win in 2009.

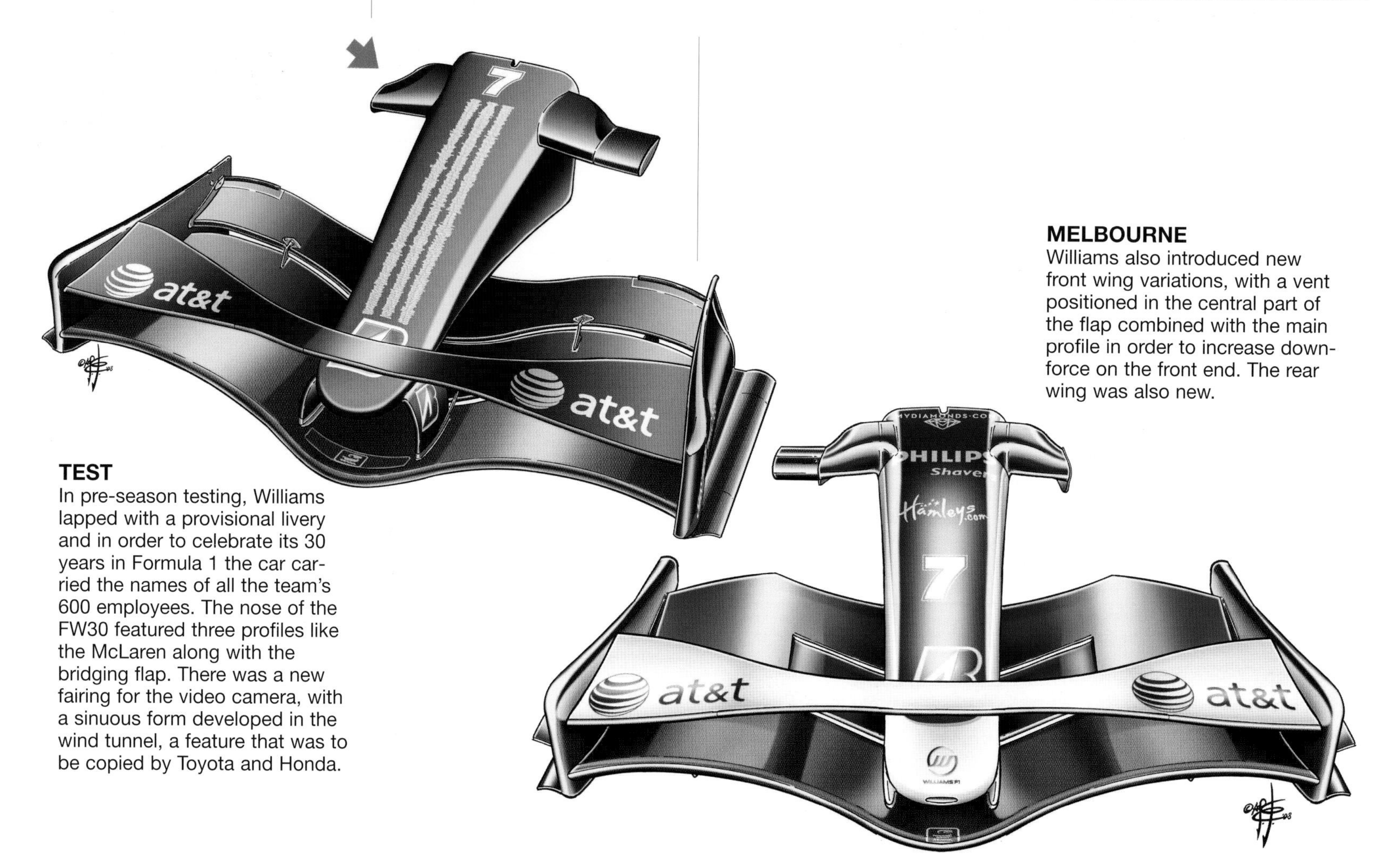

MELBOURNE
Williams also introduced new front wing variations, with a vent positioned in the central part of the flap combined with the main profile in order to increase downforce on the front end. The rear wing was also new.

TEST
In pre-season testing, Williams lapped with a provisional livery and in order to celebrate its 30 years in Formula 1 the car carried the names of all the team's 600 employees. The nose of the FW30 featured three profiles like the McLaren along with the bridging flap. There was a new fairing for the video camera, with a sinuous form developed in the wind tunnel, a feature that was to be copied by Toyota and Honda.

BRAKE AIR INTAKES

Williams' brake cooling air intakes were very sophisticated and represented an evolution of the concepts already expressed the previous year. The small oval intake served to channel air directly to the caliper, as can clearly be seen in the drawing with the drum fairing removed; this was perhaps the most complex brake air intake, with careful fluid dynamics research attempting to disturb as little as possible the flow of air in the channel between the chassis and the inside of the wheel and to expel the air outside the latter.

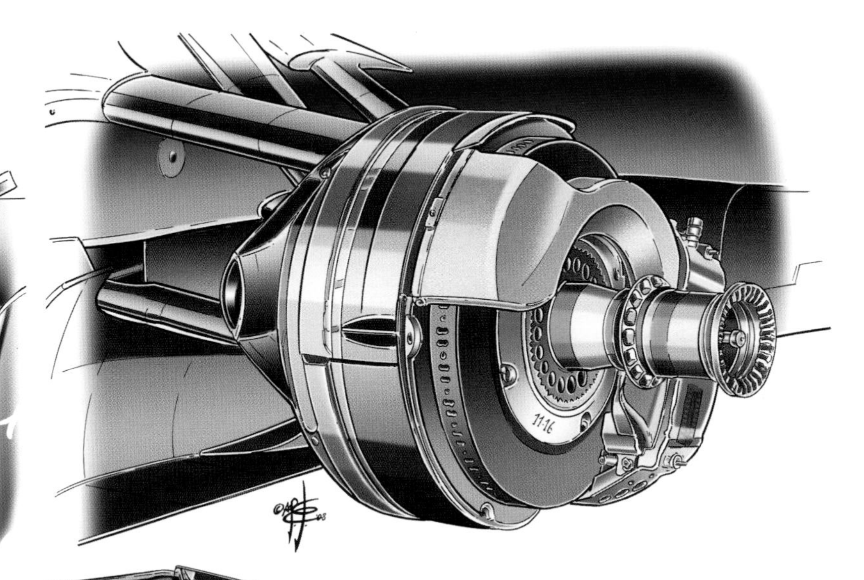

SAKHIR

Williams had two rear wings available, the low downforce version in the drawing (intended for Montreal) and the one offering a higher loading already used in Melbourne; the later was eventually chosen for both qualifying and the race and was characterised by an upwards curving centre section.

SEPANG

Williams used this new wing in Malaysia on the Friday but abandoned it for qualifying and the race. It provided greater loading and was characterised by a slightly raised central section so as to be able to use a dished profile with slightly less downforce not only in order to provide a higher top speed but also to allow the rear tyres which were suffering from the new asphalt to get up to temperature.

BARCELONA

A new front wing for the FW30, inspired by that of the BMW. It had a more deeply and more sharply dished central section and above all a main profile with a very shallow chord and a raised leading edge. The first of the two flaps had a deeper chord and acted as a support for the vertical nose-cone pylons. This meant that the team had to retake the crash test.

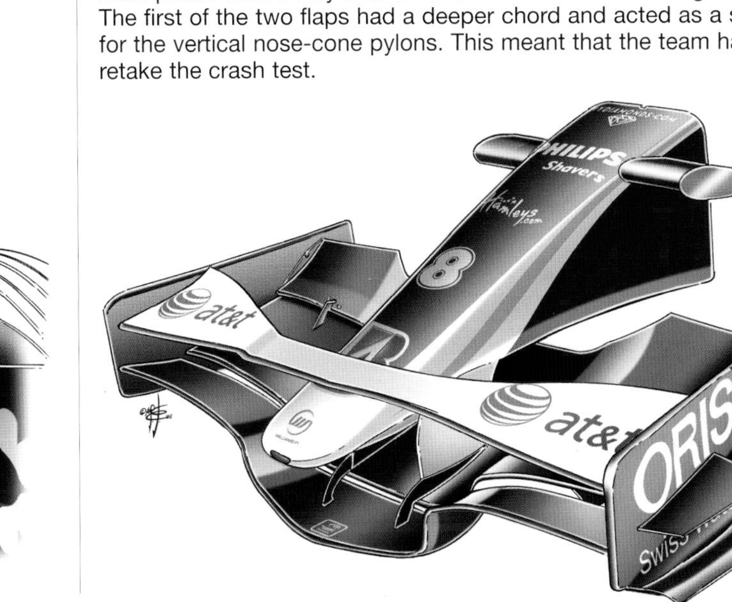

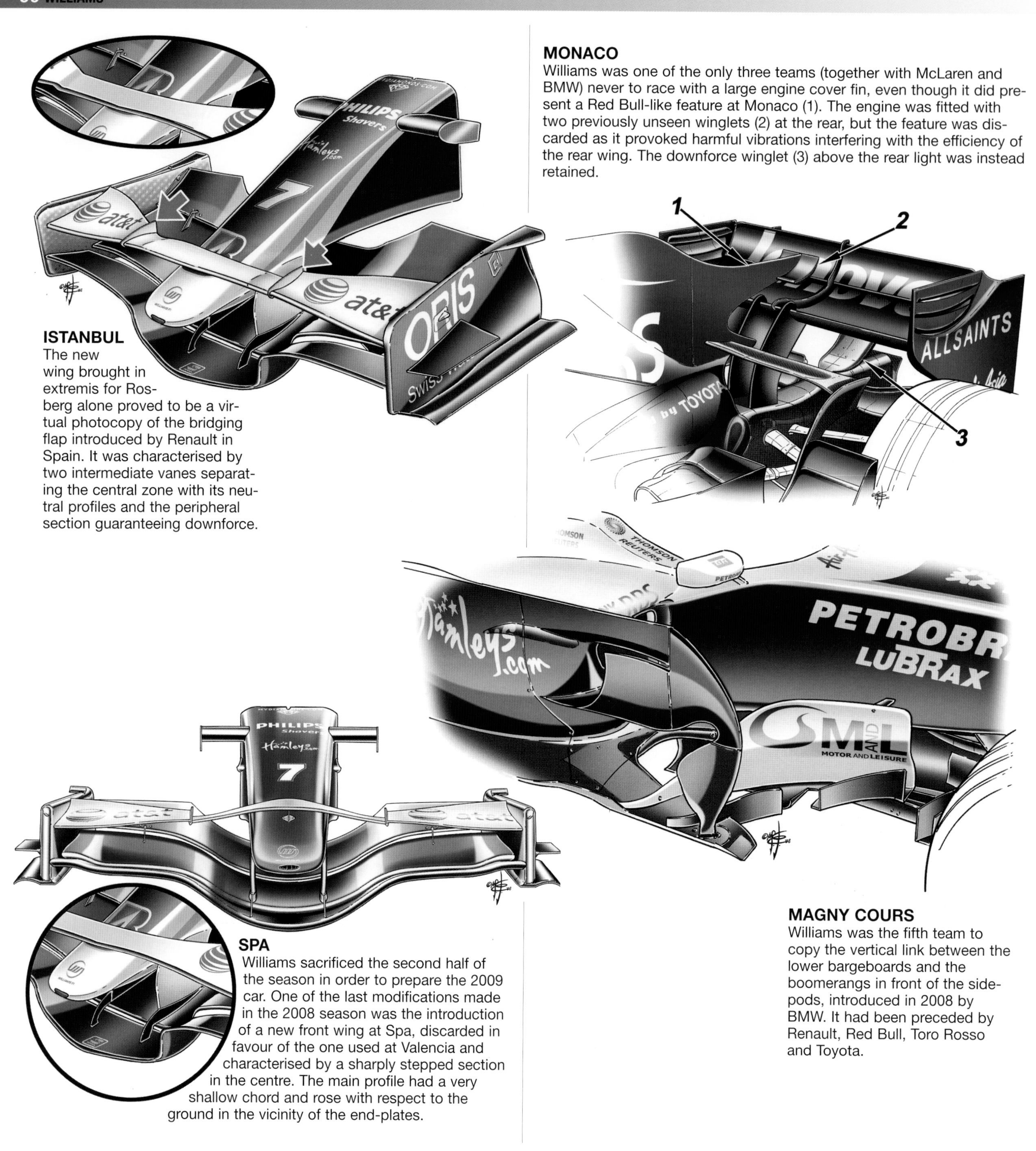

MONACO

Williams was one of the only three teams (together with McLaren and BMW) never to race with a large engine cover fin, even though it did present a Red Bull-like feature at Monaco (1). The engine was fitted with two previously unseen winglets (2) at the rear, but the feature was discarded as it provoked harmful vibrations interfering with the efficiency of the rear wing. The downforce winglet (3) above the rear light was instead retained.

ISTANBUL

The new wing brought in extremis for Rosberg alone proved to be a virtual photocopy of the bridging flap introduced by Renault in Spain. It was characterised by two intermediate vanes separating the central zone with its neutral profiles and the peripheral section guaranteeing downforce.

SPA

Williams sacrificed the second half of the season in order to prepare the 2009 car. One of the last modifications made in the 2008 season was the introduction of a new front wing at Spa, discarded in favour of the one used at Valencia and characterised by a sharply stepped section in the centre. The main profile had a very shallow chord and rose with respect to the ground in the vicinity of the end-plates.

MAGNY COURS

Williams was the fifth team to copy the vertical link between the lower bargeboards and the boomerangs in front of the sidepods, introduced in 2008 by BMW. It had been preceded by Renault, Red Bull, Toro Rosso and Toyota.

CONSTRUCTORS' CLASSIFICATION	2007	2008	
Position	8°	9°	-1 ▼
Points	6	14	+8 ▲

Another disastrous season for the Japanese giant, which dropped yet one more place in the classification despite Rubens Barrichello's fantastic third at Silverstone, the result of a perfect race strategy by Ross Brawn. The ex-technical director of Ferrari arrived at Honda when the imposition of the 2008 car had already been completed, and his experience was of little use in an attempt to make the project, born without any precise technical direction, more competitive especially after the loss of aerodynamicist Willy Toet, who moved to BMW and was replaced by Loic Bigois.
Compared to the RA107, the new Honda had a longer wheelbase to move the front axle further ahead of the sidepods to have more space to put barge boards in place of the small double elements.
The first tests failed and it was only on the eve of the Melbourne seasonal opener that a new aerodynamic package improved the situation, making the car about two seconds faster and placing the Hondas in the middle of the grid; a position that gradually slipped through their fingers as the season wore on, because the development of the RA108 was unable to achieve the rhythm of the competition.
The same aerodynamic development, with the exception of the large gullwing fins brought in on the nose and then copied by McLaren, only saw the application of elements already used by other teams (BMW's vertical fins, a "sail" on the engine cover like Red Bull) without a real logical theme.
It should be revealed that when Ross Brawn saw the impossibility of creating miracles, he aimed everything – together with the ex-

Honda RA107

Honda RA108
Launch

Honda RA108
Melbourne

Honda RA108
Barcelona

Honda RA108
Istanbul

aerodynamicist of Super Aguri, Ben Wood – at the development of the 2009 car on the basis of an aerodynamic revolution imposed by the new regulation, halting most of that which should have been the development of the season. Nevertheless, at Budapest a new rear suspension was introduced and that much improved the car's handling, contributing to giving the drivers more confidence in the car with Button, in particular, who criticised the anomalous handling of the RA108 more than his team mate Barrichello.

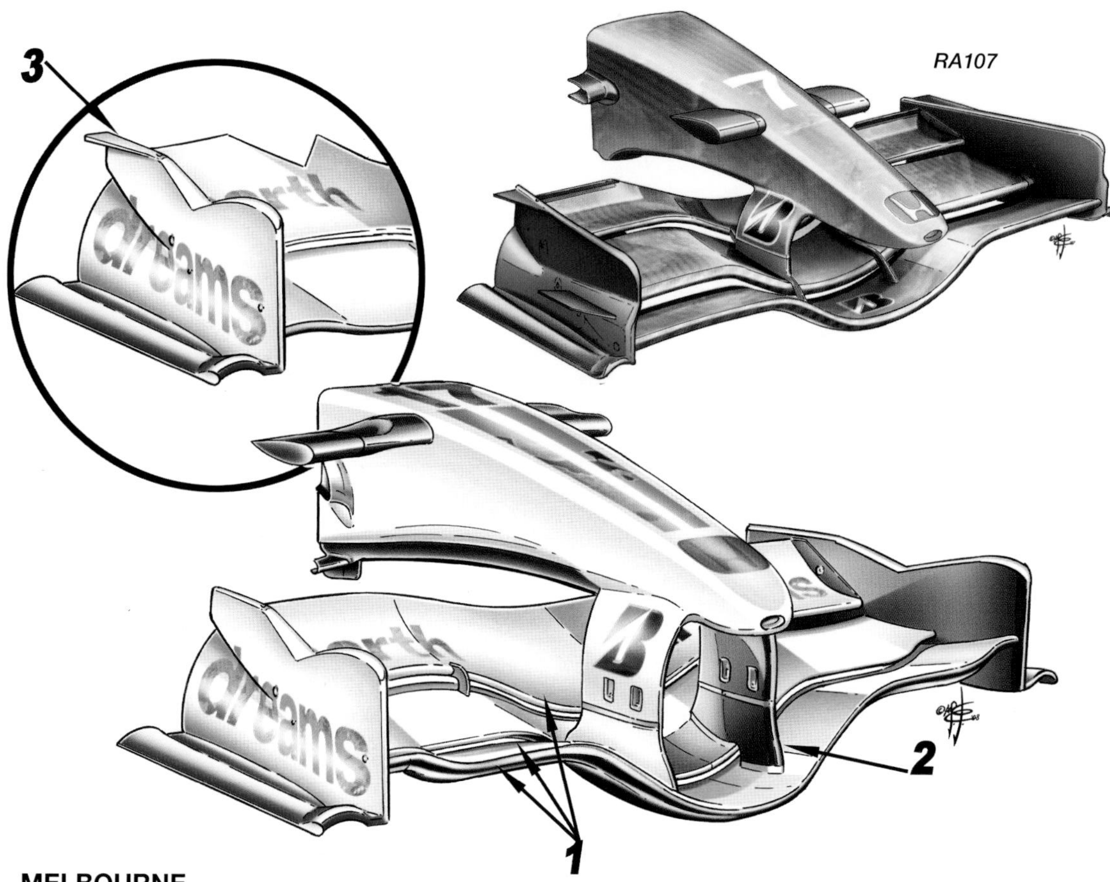

BARGE BOARDS

The RA108 dropped the double small turning vanes between the front wheels and the sidepods to adopt these very large, stepped barge boards for the purpose of reducing the harmful effect of turbulence created by the front wheels. The large boomerang of the RA106 and RA107 (see profile) was replaced with smaller units.

MELBOURNE

The Honda fielded in Melbourne was completely transformed, with new aerodynamics in all sectors. Obviously, the most visible feature was the new front wing with three elements (1), central pillars fixed (2) mid-way between the main plane and the first of the two flaps. The end plates were also slightly modified at the trailing edge, which had no triangular finlets (3), although they were present on the old two-plane wing.

ISTANBUL

Honda set aside the almost closed wheel fairing introduced in Spain, which overheated the front brakes too much. It used a similar cut with the air vent, inclined in a like manner to that of Ferrari.

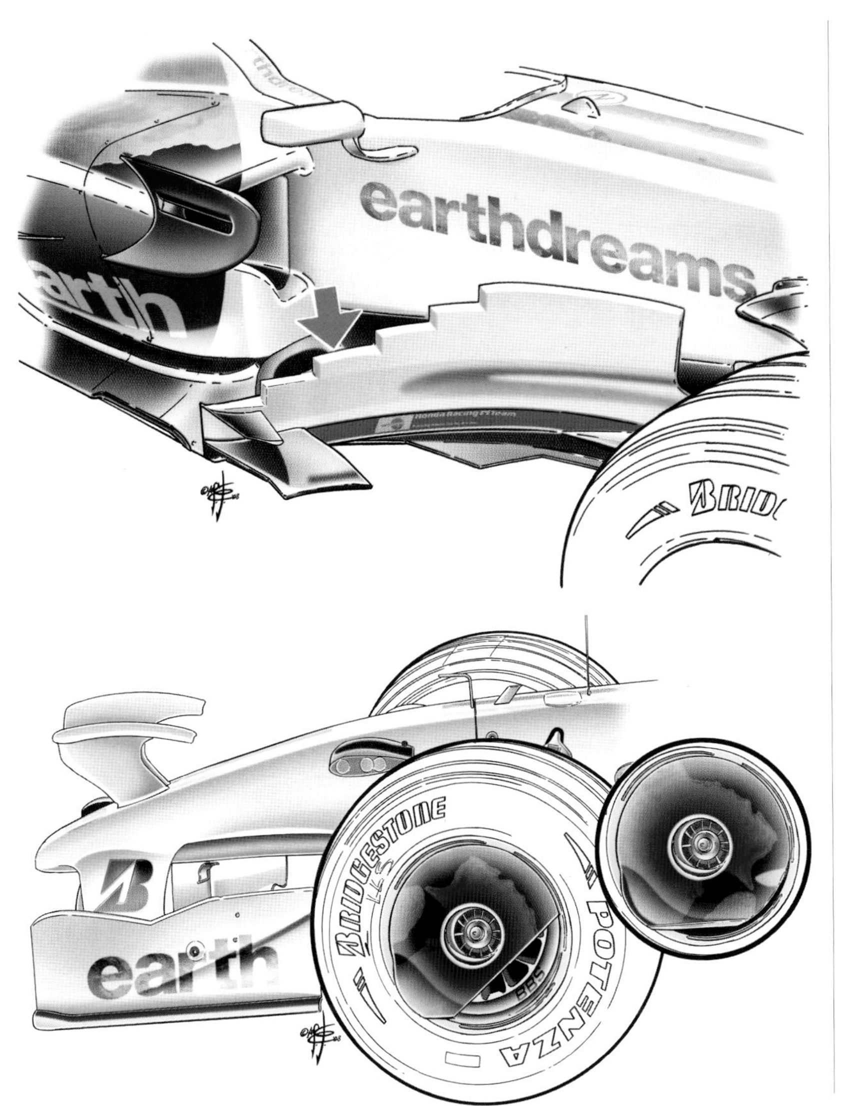

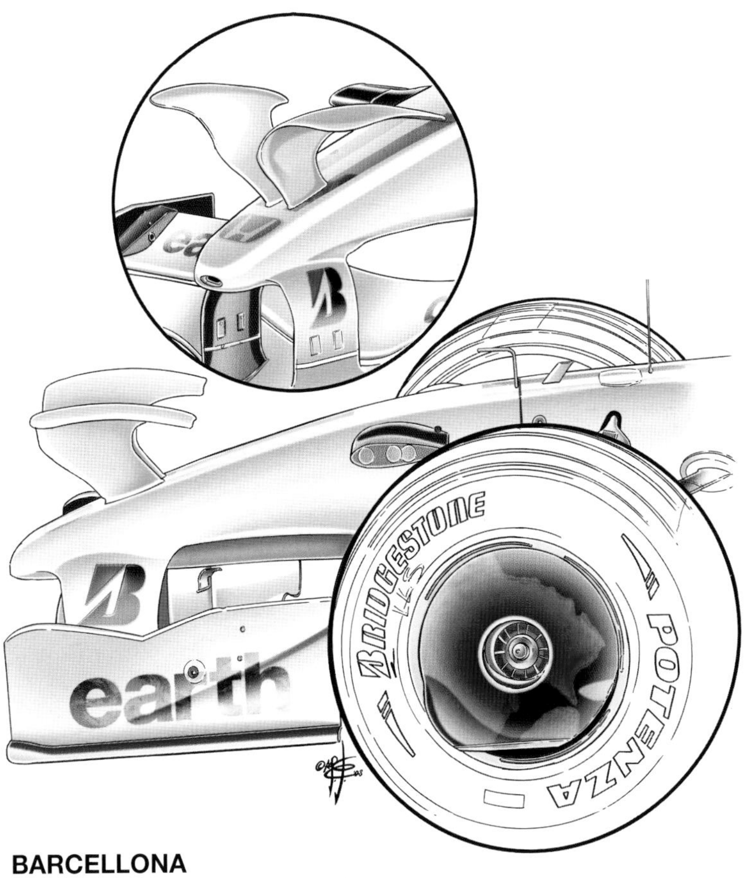

BARCELLONA
The RA108 brought back a solution tried in the 2007 pre-season tests but never used in a GP; these curious big gullwings mounted on the nose to redirect the air flow and avoid the downforce of the central part of the nose. The front wheel fairing was tried again on the Friday morning, but was not raced as it created some overheating problems.
The twisted rear wing was only tried by Barrichello but was dropped.

REAR WING

A new rear wing for the Hondas, based on the one used in Canada in 2006 (see circle), which integrated the separate Renault style flap with the very sinuous and sophisticated shape of the planes. Barrichello ran a comparison test with the new plane and a high download version, but in the race both drivers used the new wing.

Montreal 2006

MONTREAL

Honda used the same rear wing as the one in Turkey, even though the flap was cut in its centre to improve top speed, which was always very low due to the limited power of the Honda engine compared to its adversaries.

MONACO

Honda brought in two front wings based on the new unit with the large gullwings introduced in Turkey. Both had barge boards (1) at the sides to increase the efficiency of these wing appendages. The finlet (in red) inside the end plates was new. The one used is shown in the illustration, with new end plates produced by uniting the previous version (straight) with the one that appeared in 2007 in Hungary (3), as can be seen from the conjunction line of the two elements. The triangular finlets (2) at the extremes of the end plates and the bigger horizontal one (4) at the beginning were new; note the outward inclination of the lower part (5) to increase the amount of air directed under the principal plane.

HOCKENHEIM

By practice for the GP of Germany, Ferrari and Honda had introduced the vertical link between the lateral boomerang and the lower barge boards to better direct the air flow toward the rear of the car, cleaning it of turbulence from the front wheels. It was a feature brought in by BMW that made converts among all the teams, with the exception of McLaren and Force India.

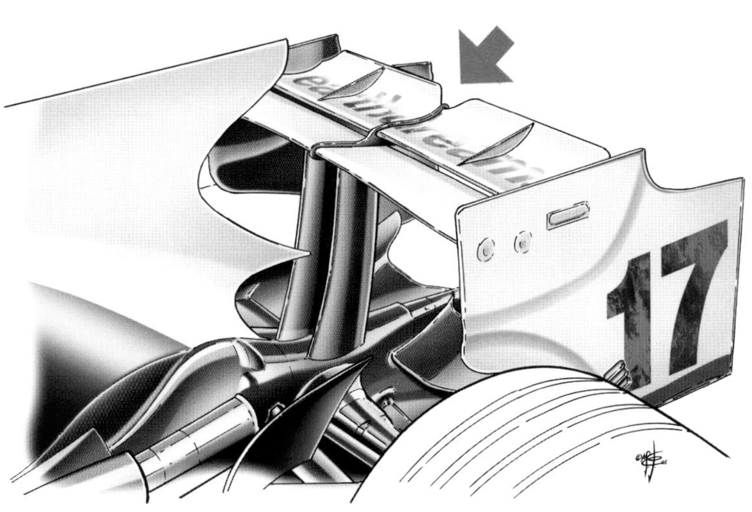

MONZA REAR WING

On Barrichello's car, Honda used a rear wing with a cut in both the area near the end plates and the centre, where there is a shadow zone originating from the large fin to reduce turbulence and improve top speed.
Button, on the other hand, raced with the Montreal wing to have more stability in the chicane.

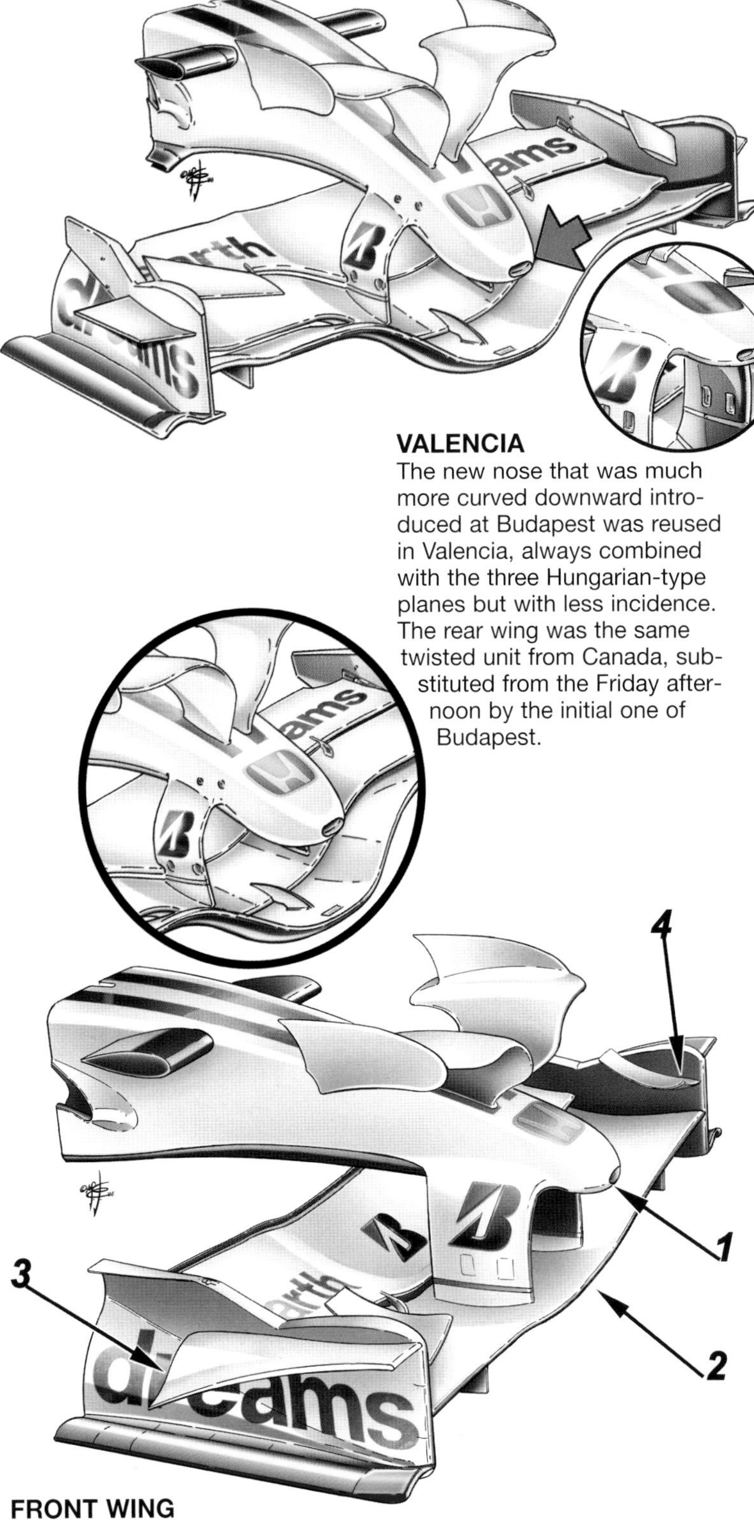

VALENCIA

The new nose that was much more curved downward introduced at Budapest was reused in Valencia, always combined with the three Hungarian-type planes but with less incidence. The rear wing was the same twisted unit from Canada, substituted from the Friday afternoon by the initial one of Budapest.

FRONT WING

Somewhat unusual was the front wing always employed at Monza by Honda and based on the nose of the start of the season, with the central part less curved downward. The shape of the keels at the outside was curious as they pointed downward and linked with the twisted internal one. The main plane dropped its accentuated spoon shape and was almost straight, coupled with a flap with a notable chord and a Gurney flap for its entire width, against the current tendency to carry out small interventions.

Another change of name for the team based at the old Jordan premises at Silverstone, even if the car still remained a Spyker in the beginning, having been derived from the 2006 Midland. In effect, for the technicians directed by Mike Gascoyne the 2008 project was based on the B version that made its debut at Monza in 2007 with a gearbox that had to last four consecutive races. Almost all the season's work was concentrated on aerodynamic research with the objective of increasing downforce. Gascoyne pushed hard in that sector, recouping, among other things, features he first brought in on the distant 1995 Tyrrell 018 with the Yamaha engine. It was the more copious fairing of the lower wishbone of the front suspension and that of the mirrors integrated in a new plane that was, in turn, connected to the lateral "boomerang". The BMW-style vertical connections that were used by almost all the 2008 cars were never introduced on the Force India. But at almost every race there were small developments, although the key moments of the season were three: at Monaco, Silverstone and Valencia.
On the narrow roads of Monte Carlo, the stage for Adrian Sutil's stupendous race in the wet, the team failed in its possible major achievement of coming fourth only because of Raikkonen's fault.
An important aerodynamic package bowed in at Silverstone, the most visible element of which was the large fin on the engine cover. There was also a new front wing, sidepods and an inertial damper.
On the new Spanish circuit, the quickshift version of the car came in at last. As in 2007, the car was powered by the 8-cylinder Ferrari engine, so that the team at least had the satisfaction of fighting on a level playing field with Honda at the end of the season, despite the incredible budget and personnel difference.
However, Mark Smith joined the team to watch over the evolution of the new gearbox.

Spyker M16

Force India
Melbourne

Force India
Sepang

Force India
Silverstone

Force India
Sao Paolo

FORCE INDIA

CONSTRUCTORS' CLASSIFICATION	2007	2008	
Position	10°	10°	= ■
Points	1	0	-1 ▼

MELBOURNE

Mike Gascoyne dusted off an idea he first introduced on the 1995 Tyrrell 018. It was fairing the lower wishbone of the front suspension with a real wing plane and exploiting to the maximum the limits permitted by the regulations of 70% of the wishbone's length.

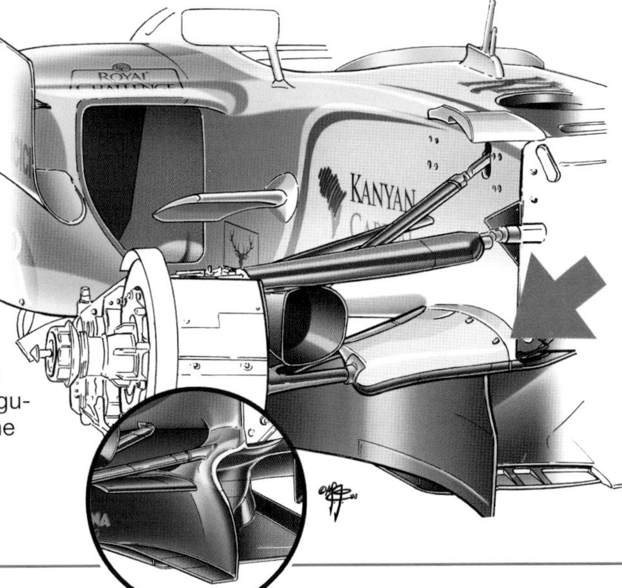

TYRRELL 018

In Formula 1, ideas are often brought back and Mike Gascoyne once again took up the idea of a fairing suspension elements in an extensive manner. In 1995, the component affected was the upper wishbone, which was completely incorporated in a single wing plane, but in 2008 the partial fairing concerned the lower wishbone. It should be pointed out that to avoid exaggeration, in 1995 the Federation banned total fairing, imposing a limit of 2.8 for the planes' chord in relation to the thickness of the wishbone levers. Today, that limit has moved on to 3.5 for the chord and no more than 70% of the length of the wishbone.

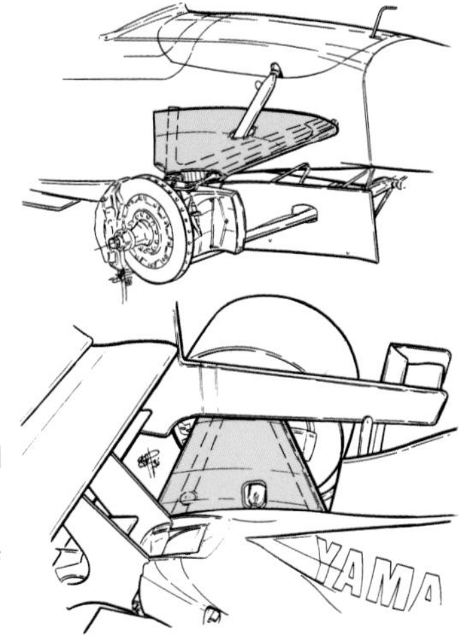

SEPANG

Mike Gascoyne introduced many new aerodynamic developments on the basis of the previous year's car. At the opening race of the season an almost total fairing of the lower wishbone of the front suspension was brought in, exploiting the maximum limit of the regulation. Immediately afterwards, the extreme fairing of the mirror supports became a true wing plane to straighten the air flow to the cockpit, as on the Ferrari.

SILVERSTONE

Force India followed the fashion of the "sail-type" engine cover launched by Red Bull and copied by Renault and Toro Rosso. The team did so at Silverstone, combining it with the new McLaren-school sidepods, in other words with the boomerang-chimney-winglet group integrated together.

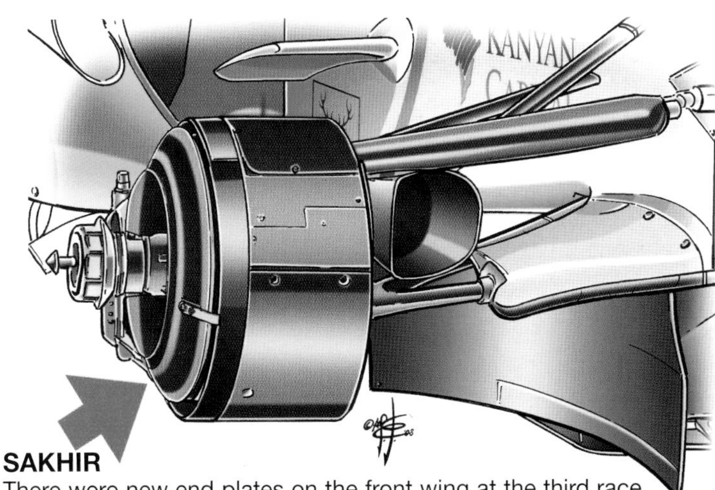

SAKHIR

There were new end plates on the front wing at the third race of the season, as well as this second drum on the external part of the front brake discs. Until the GP of Malaysia, the cover arrived at the internal wire of the Hitco discs.

HOCKENHEIM

There was further integration of the mirror supports, which had become real curled wing planes that joined up with the end plates, avoiding and reducing the downforce effect of the body in that area.

The adventure of Super Aguri, the youngest team in the Formula 1 World Championship – it made its debut in 2006 – lasted only four races, despite showing some good results in the 2007 season, scoring four championship points. As with the previous year, the starting point was the Honda RA107 engine and gearbox, which were adapted to last four races as required by the new regulations. The group, directed by Mark Preston, did good work to improve the cooling system and modify the chassis in the rollbar area with the objective of passing the new side crash test at the height of the driver's head, as a result of which its weight was reduced and all but cancelled out the chance of moving its ballast around. Aerodynamic development was by Ben Wood, who later moved on to Honda with the break-up of the team after those four races and the GP of Spain. The SA08 was given a single, simple test drive before the beginning of the season, achieving in the end no fewer than 5 finishes out of 8 starts; a reasonably positive result, given the conditions.
Unfortunately, there was a complete lack of financial support from Honda headquarters, among others, which more or less limited itself to supplying engines and gearboxes.
Super Aguri retained the Alcon braking equipment, supplied exclusively to Honda in 2007.
Compared to the Honda RA107, the worst car produced by the Japanese team in recent years, new developments concerned the cooling system, the sidepods, the diffuser and the rear wing end plates.
In Spain, given the difficulties with the sponsors, Super Aguri raced with a different livery and without many logos. At the fifth race in Turkey, the team's material was not even unloaded from the trucks, because their last hope had disappeared so the brief adventure of this team in Formula 1 finished forever.
This also marked the end of the Formula 1 career of the 2002 Arrows A23 which had been exhumed in 2006. Such an enterprise had seemed impossible but

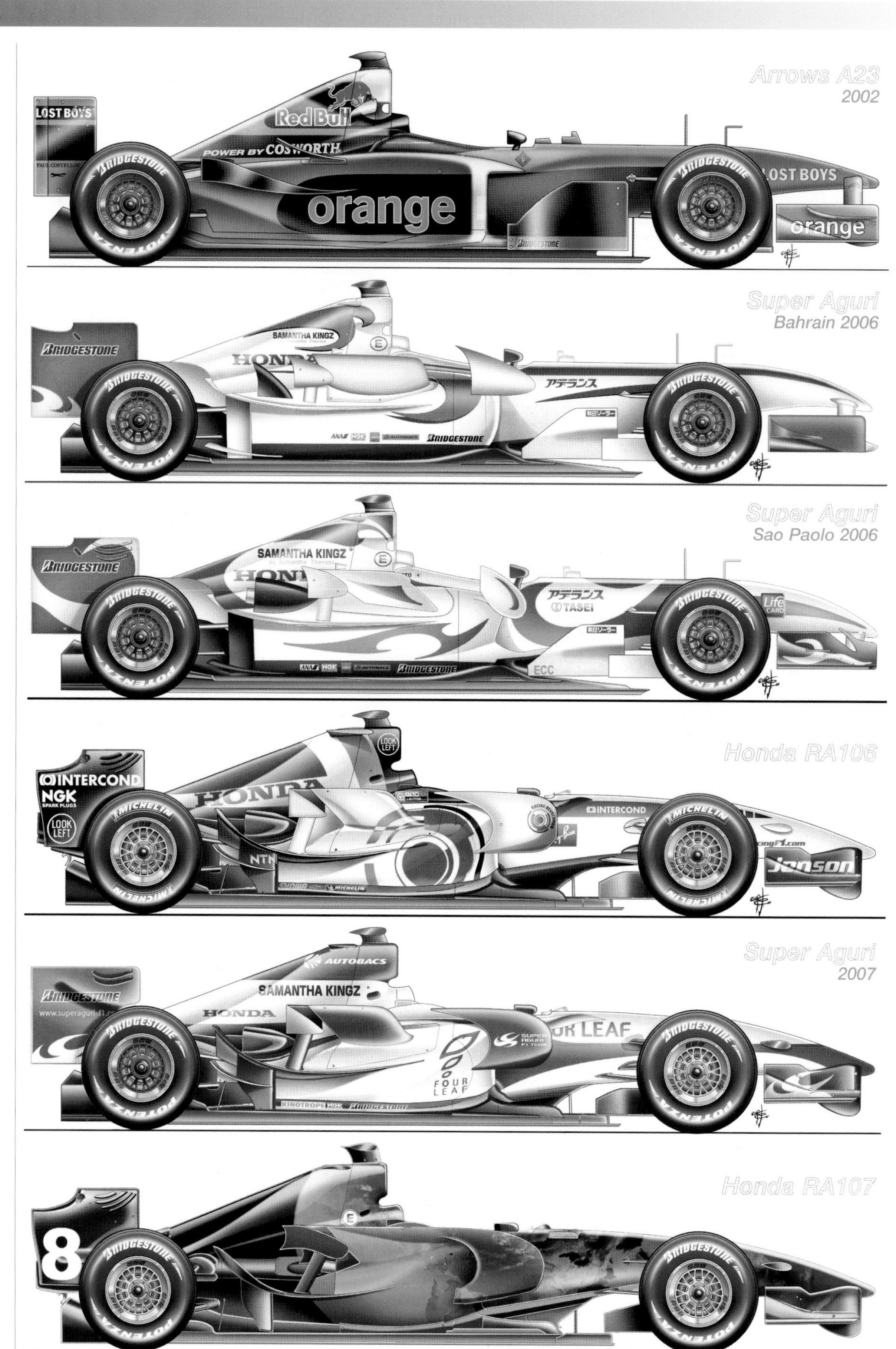

SUPER AGURI

CONSTRUCTORS' CLASSIFICATION	2007	2008	
Position	9°	11°	-2 ▼
Points	4	0	-4 ▼

Super Aguri
Melbourne 2008

Super Aguri
Barcelona 2008

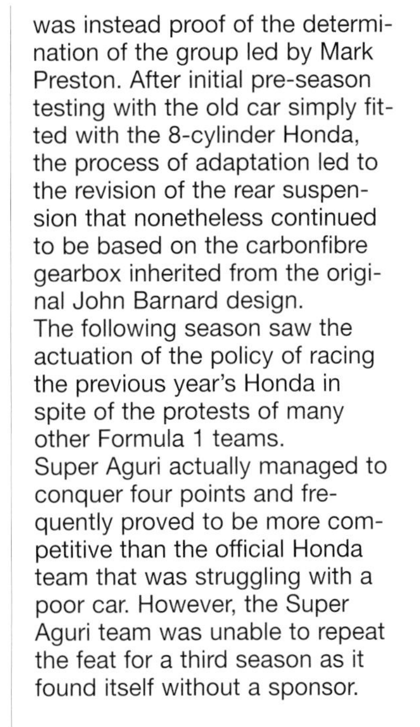

was instead proof of the determination of the group led by Mark Preston. After initial pre-season testing with the old car simply fitted with the 8-cylinder Honda, the process of adaptation led to the revision of the rear suspension that nonetheless continued to be based on the carbonfibre gearbox inherited from the original John Barnard design.
The following season saw the actuation of the policy of racing the previous year's Honda in spite of the protests of many other Formula 1 teams.
Super Aguri actually managed to conquer four points and frequently proved to be more competitive than the official Honda team that was struggling with a poor car. However, the Super Aguri team was unable to repeat the feat for a third season as it found itself without a sponsor.

FRONT WING

Super Aguri went to the first Grand Prix in Melbourne with a front wing strongly inspired by that of the mid-season Honda. All the aerodynamic updates were applied without any wind tunnel testing by Ben Wood, who moved on to Honda with Ross Brawn to work on the 2009 Honda project.

REAR SUSPENSION

The rear suspension was totally new compared with the old Arrows. The biggest difference concerned the unusual position of the brake calipers which are usually placed behind the axle as on the old car, but Mark Preston moved them high up in front of the rear axle. The new installation of the various suspension components was much cleaner and more compact.

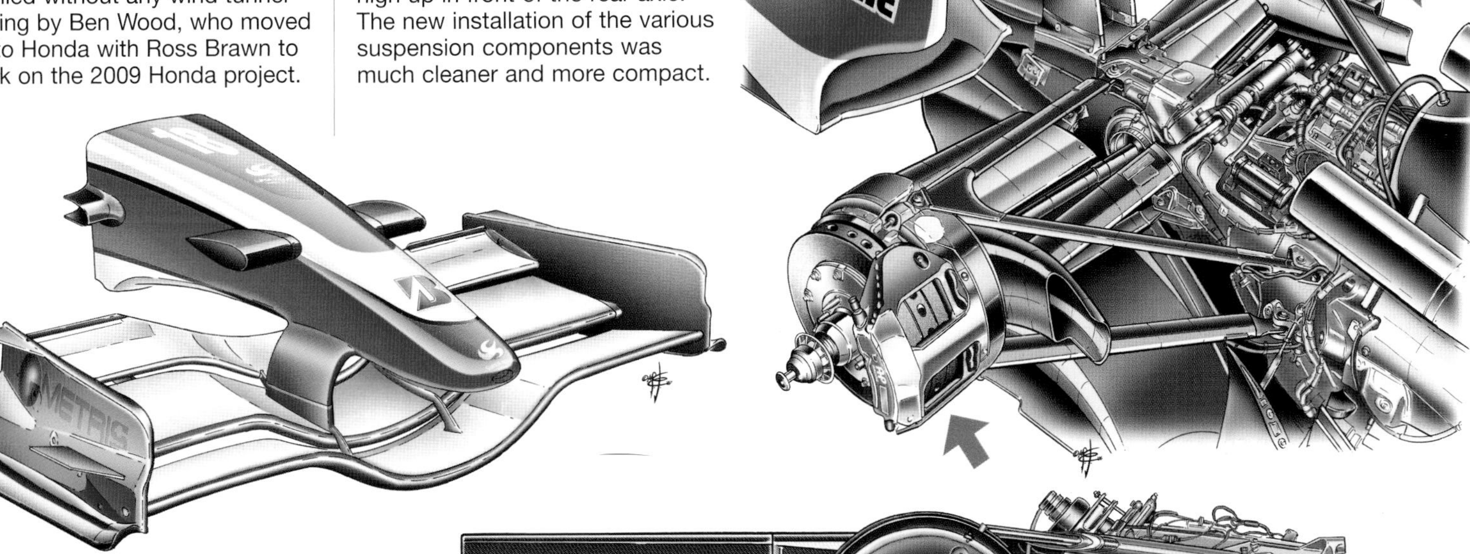

GEARBOX

Super Aguri received from Honda the 8-cylinder engine and gearbox with the 'box itself in carbon fibre. The internal mechanisms were revised to last the four Grands Prix required by the new regulations. As far as the car was concerned, in the side view above the protection at the sides for the driver's head one can see the different shape of the fin-chimney group compared to the 2007 Honda.

Ferrari F60

McLaren MP4-24

BMW F1 09

Renault R29

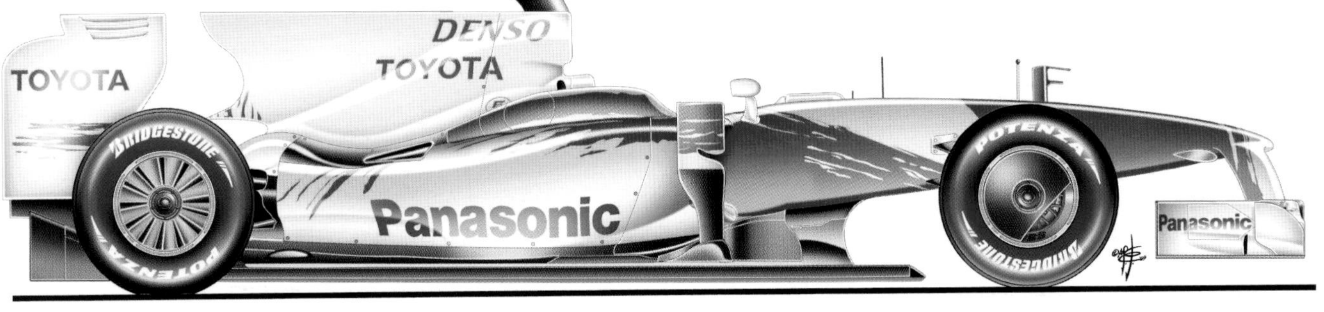

Toyota TF109

The 2009 SEASON

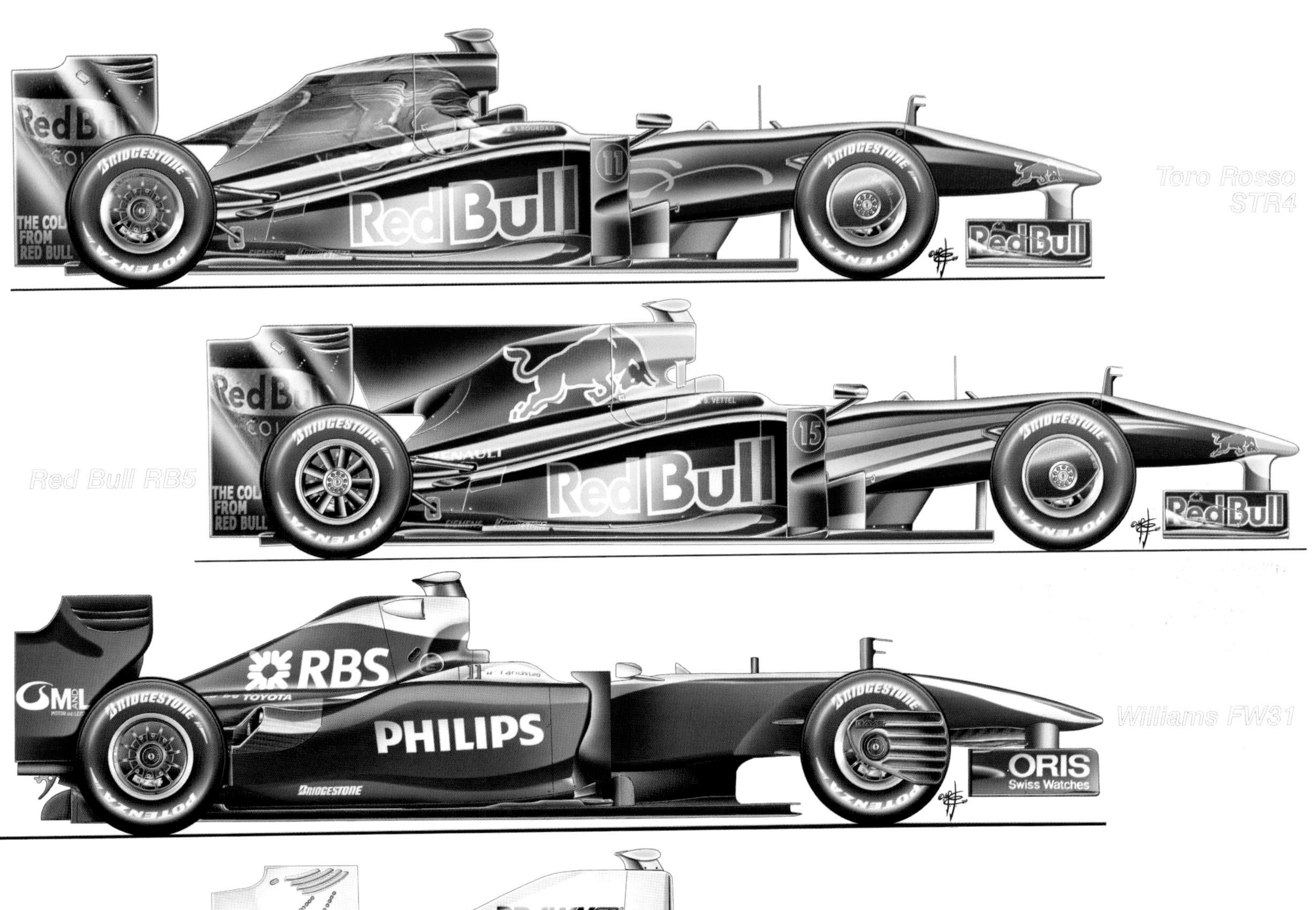

Toro Rosso STR4

Red Bull RB5

Williams FW31

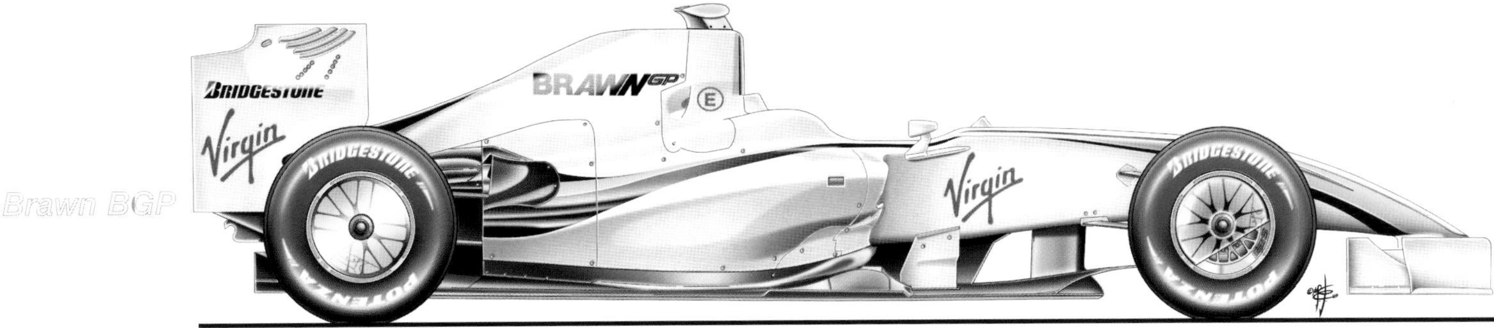

Brawn BGP

Force India

The new technical regulations introduced by the Federation for the 2009 season comprise, without doubt, the most devastating revolution in the history of modern Formula 1. Not a single sector of the car was immune from severe modifications, to the extent that even a layman would have been surprised when observing the new shapes imposed on the cars. Shapes, among other things, that should have left little room for imagination even if, with the presentation of the new cars, we saw that unfortunately the text of the regulations left so many grey areas and that they were 'exploited' by the various designers to create cars that were equally as sophisticated as their predecessors. In these pages we illustrate the fundamental components of new regulations, leaving the task of analysing that which later became the constantly nagging nuisance of the start of the year left to the chapter that introduces the 2009 season with the interpretation made by three teams of one hot potato of a subject – diffusers. Here we analyse, therefore, the text of the new regulations and their principal inspirers. At the basis of everything was the attempt to bring to life a car that felt less of the negative effects of slipstreaming in order to make overtaking easier, even if in the last two seasons such manoeuvres on the track were seen – and how!
They also wanted to stop wind tunnel research, even if that kind of development had never been used more intensely than in preparation for the 2009 season; so much so that no fewer than three teams, Ferrari, McLaren and Renault, went to America to test in the scale 1/1 wind tunnel belonging to Windshear in Charlotte, North Carolina. It is, in fact, more than obvious that the objective of the Federation was to reduce the downforce capacity by 50% with the elimination of all those aerodynamic appendices that had covered the cars in recent seasons were skilfully reduced by the imagination and skill of the technical offices of the various teams and all of this, of course, at notable cost, as was the KERS project.
But in this chapter, we shall concentrate on the cuts made to aerodynamics and on how they were supposed to have changed the cars.
In theory, away with all the aerodynamic appendages inside and behind the front wheels, away with the chimneys, the boomerangs and anything else that was applied to the sides, which had to be shaped to extremely precise radii. Obviously, the mini-planes behind the engine air intake were banned as well as the one 100 cm wide inside the rear wheels. The side vents to evacuate hot air from the sidepods were also abolished; a limitation that created substantial difficulty for the designers, aggravated by the further need to cool the KERS batteries. The strange dimensions of the wings were dictated by the need to reduce the negative effects of the vortices created by the car when overtaking.
The front wing, very wide at 1800 mm, was lowered by no less than 75 mm and 'works' in a less disturbed zone. It must have a neutral plane and be the same for all the cars in the central 500 mm area, where the actual wings had a generous spooned zone.
In addition, there was the new rule introducing the flap that could be regulated (6°) by the driver. The rear wing is narrow but raised 150 mm in order to have a less disturbed air flow and a greater – even if slight – possibility of incidence of the planes.

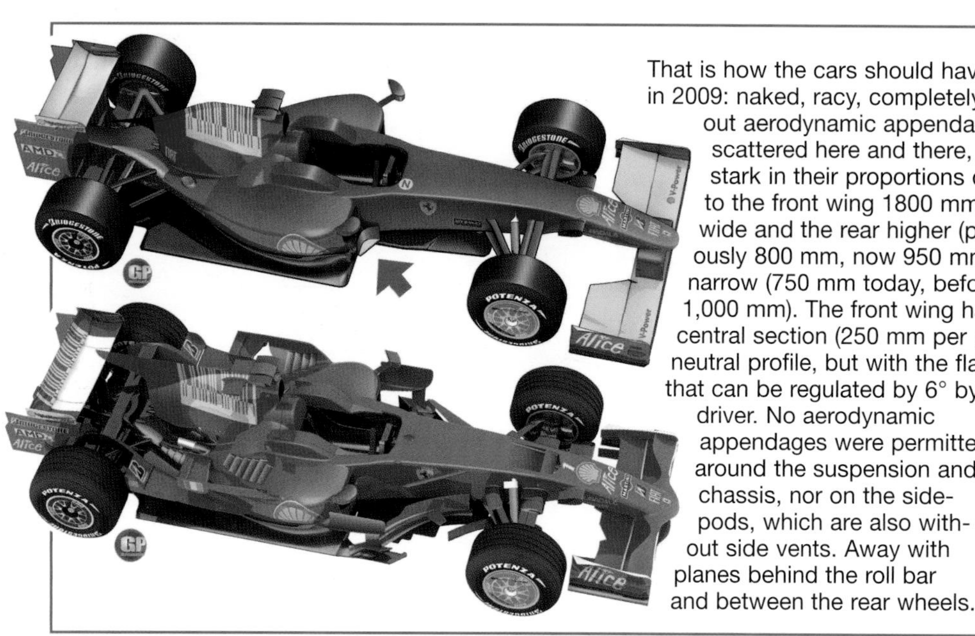

That is how the cars should have been in 2009: naked, racy, completely without aerodynamic appendages, scattered here and there, but stark in their proportions due to the front wing 1800 mm wide and the rear higher (previously 800 mm, now 950 mm) and narrow (750 mm today, before 1,000 mm). The front wing has a central section (250 mm per part) of neutral profile, but with the flaps (3) that can be regulated by 6° by the driver. No aerodynamic appendages were permitted around the suspension and chassis, nor on the sidepods, which are also without side vents. Away with planes behind the roll bar and between the rear wheels.

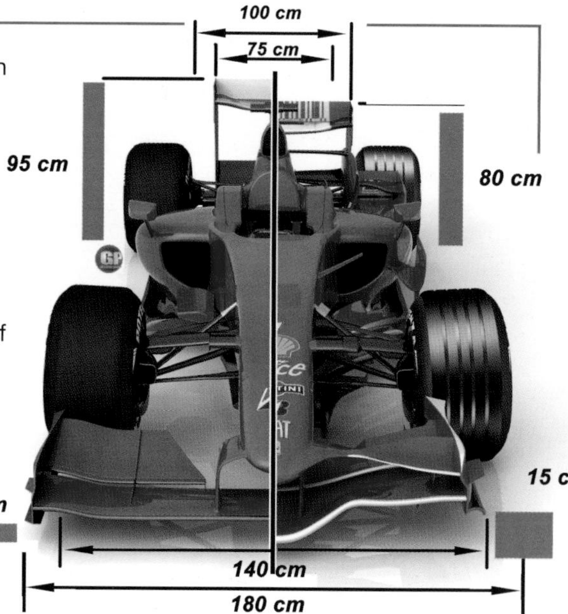

SLIPSTREAMING EFFECT

As raced in 2008, the cars had aerodynamics integrated among each other that were so sophisticated that they were too sensitive to external elements, as in the case of the slipstreaming effect; from that it was necessary to develop a new technical regulation that was able to reduce the negative effects of this phenomenon. Not by chance, the slipstreaming effect began to make itself felt, as seen in the graphic, at seven lengths from the car up front with a reduction of load at the rear end, then at four cars' lengths it compromised the engine's fuel feed. But worse still, at one length's distance it almost totally eliminated the downforce at the front end. The car suddenly became an understeerer, making the overtaking manoeuvre difficult. The loss of grip at the front influenced not only the car's directionality but also it's braking, notably reducing the brakes' precision and effectiveness.

TOP VIEW

In the view from above, the different dimensions of the front wing (1) are even more evident, especially in relation to the front tyres. Also indicated is the portion of the flap (2) that can be regulated. There were also limitations to the suspension fairing, which can no longer have the section near the chassis made bigger with the wing planes.
The 3.5 ratio between the width of the planes and thickness of the wishbones remained. (3) There were no more turning vanes and fins of various types in the zone between the front suspension and the sidepods: there is just a small triangular portion that is 'free', indicated by the red arrow where we can see a few aerodynamic appendages. (4) No more chimneys, but above all no more side vents to dissipate heat from the sidepods. The mini-planes at the sides of the roll bar were obviously abolished. (5) The only permitted apertures are for the exhaust pipes and the suspension arms. (6) Also banned is the wide plane (100 cm) with reduced chord placed in the area of the rear axle. And then the body has to have constant radii and no longer include any type of angle or fin stumps.
The reduction in the width of the rear wing is clear to see.

FRONT VIEW

From the frontal view one can see well how the front plane is much wider (it created problems at the start and in 'hand-to-hand' fighting on the track) and has the flap that can be regulated by the driver. The modifications to the diffuser plane are very complicated and are shown in a lateral comparison. Here one can see that in practice the wing channels have been moved beyond the rear axle; from the rear view, one notes how the central ones, previously very high, are now at the same height (increased by 50 mm compared to 2008) as the two laterals.

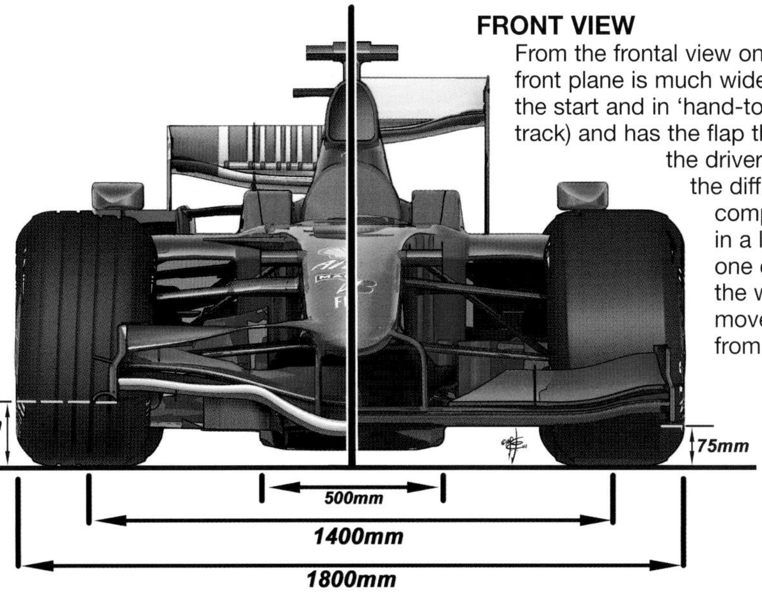

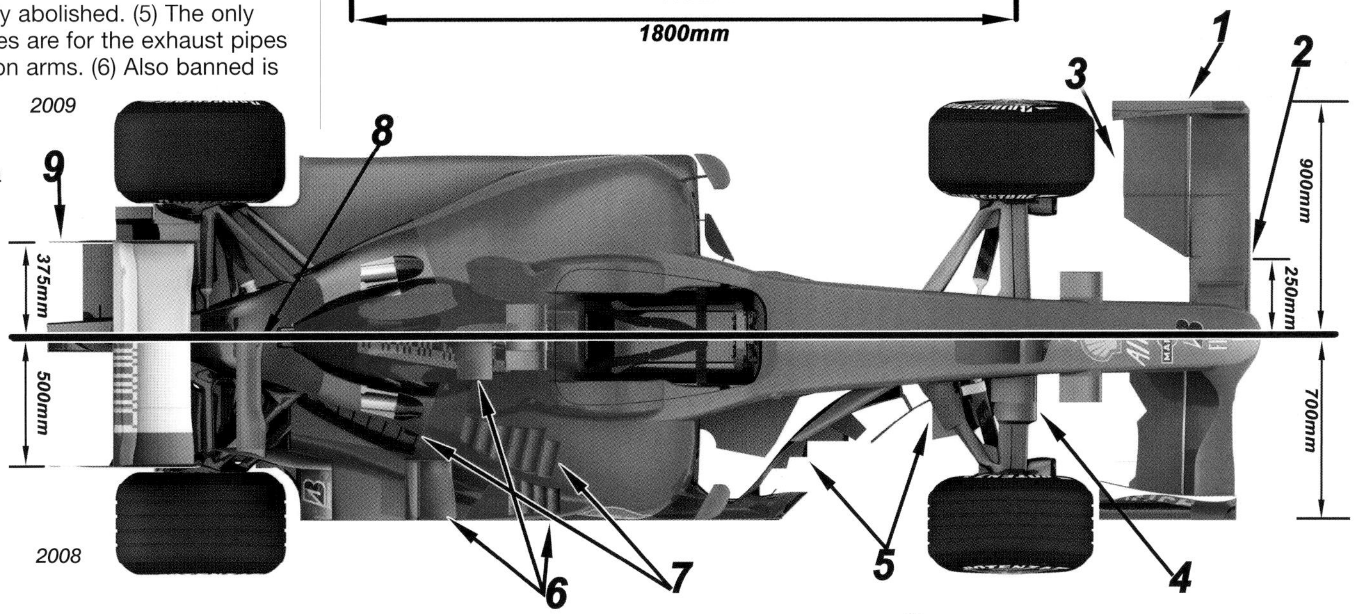

MOBILE WING

Once again to help facilitate overtaking, for the first time after mobile wings were banned in 1969, the possibility of regulating the flaps (6°) was conceded (see the design at the side). This operation is permitted twice per lap: once to modify the incidence of the flaps and the other to return to the starting position. In the illustration below we show the mini-plane positioned behind the roll bar, first introduced by Ferrari at the 1968 GP of Belgium, which was at the centre of mistaken use by Jacky Ickx at the GP of Italy.

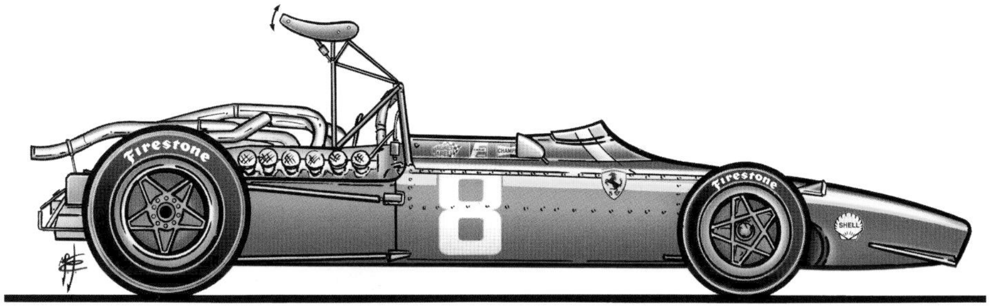

REAR WING AND DIFFUSERS

The rear wing of narrowed width is notably raised to the level of the roll bar, from 800 mm to 950 mm from the reference plane. To compensate for the reduction in width, the possibility of incidence was raised by 2 cm, from 200 to 220 mm. There are severe modifications to the diffusers, the central one of which was heavily limited and must now have the same height as the lateral units. The latter have been moved back and raised: previously, they started at almost the height of the front tangent of the rear wheel and they halted at the rear axle, while the central unit continued. But for 2009 all three channels begin level with the rear axle and finish with an overhang of 350 mm. The height has been increased to 175 mm but the same for all the channels while in 2008 the central diffuser was heigher. The principal difference is that they have been moved back in relation to the rear axle, as can be easily seen in the diffusers' design of comparison. To compensate for these aerodynamic cuts and the reduced downforce, the Federation has reintroduced slick tyres, which do so by returning about 20% more mechanical grip.

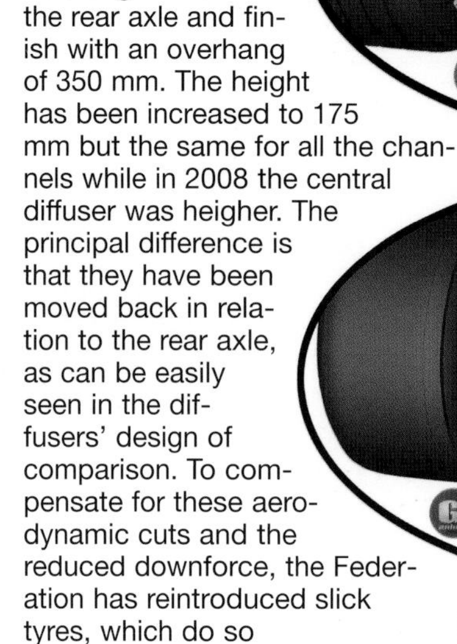

2008

2009

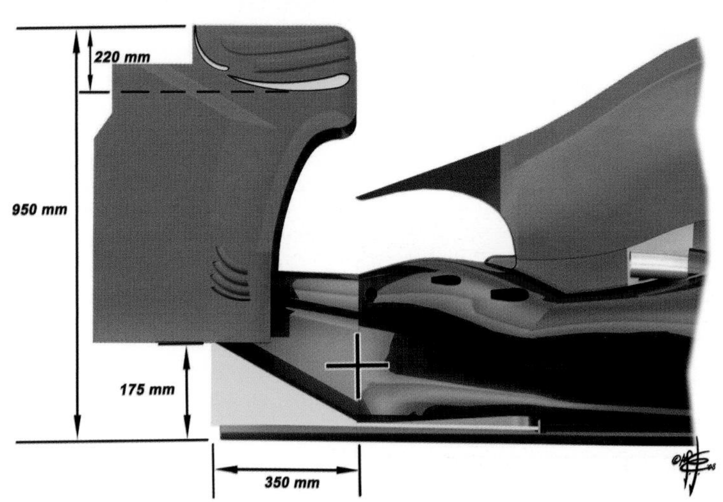

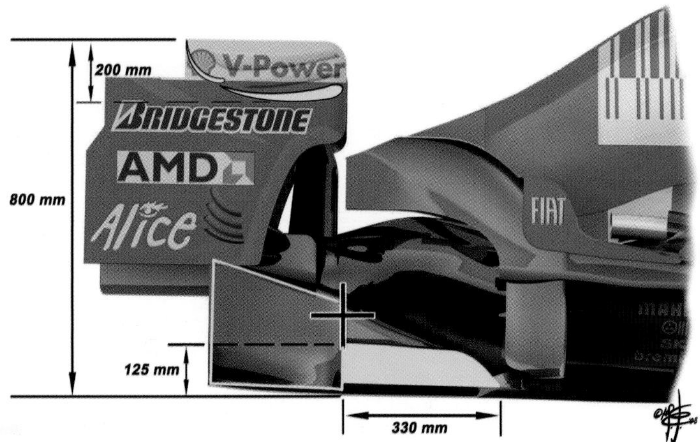

F2008

OVERTAKING PROBLEM

To ensure a good load on the front end, the Federation has increased the width of the front wing to a maximum, taking it to the external measurement of the car of 180 cm. Doing so increases the bulk of the car in an incredible manner, with undoubted negative effect at the start of a race and while overtaking in slow corners. As can be seen in the illustration, the space to slip in between one car and another at the start is, as a result, severely penalised. We have simulated the serpentine between two cars at the start, where it can be immediately seen that in 2009 more space is needed to pass. That risks, obviously, the breakage of the front wings, the bulk of which must be evaluated blindly by the drivers, given that its extremities are covered by the front wheels.

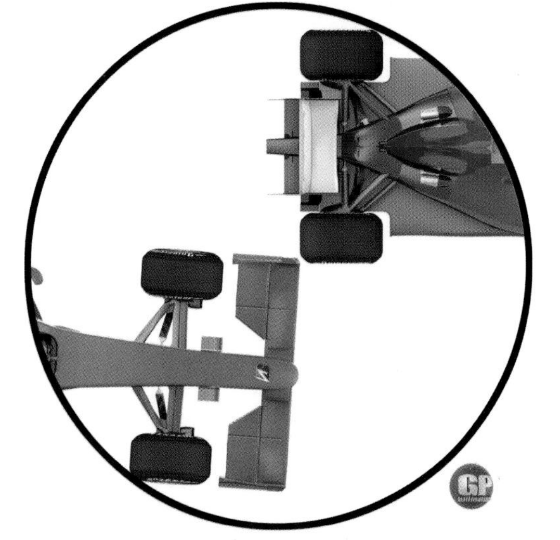

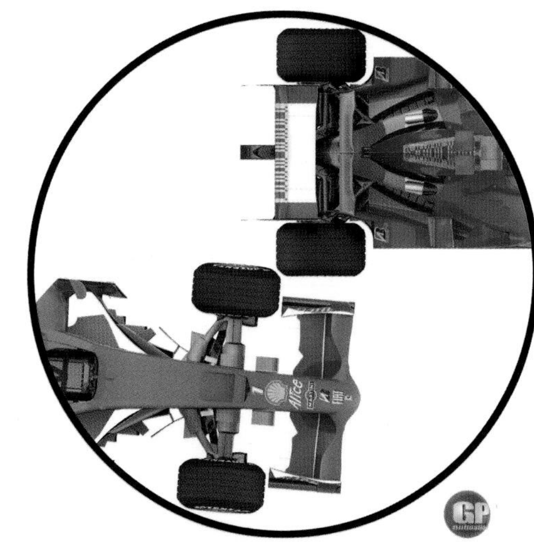

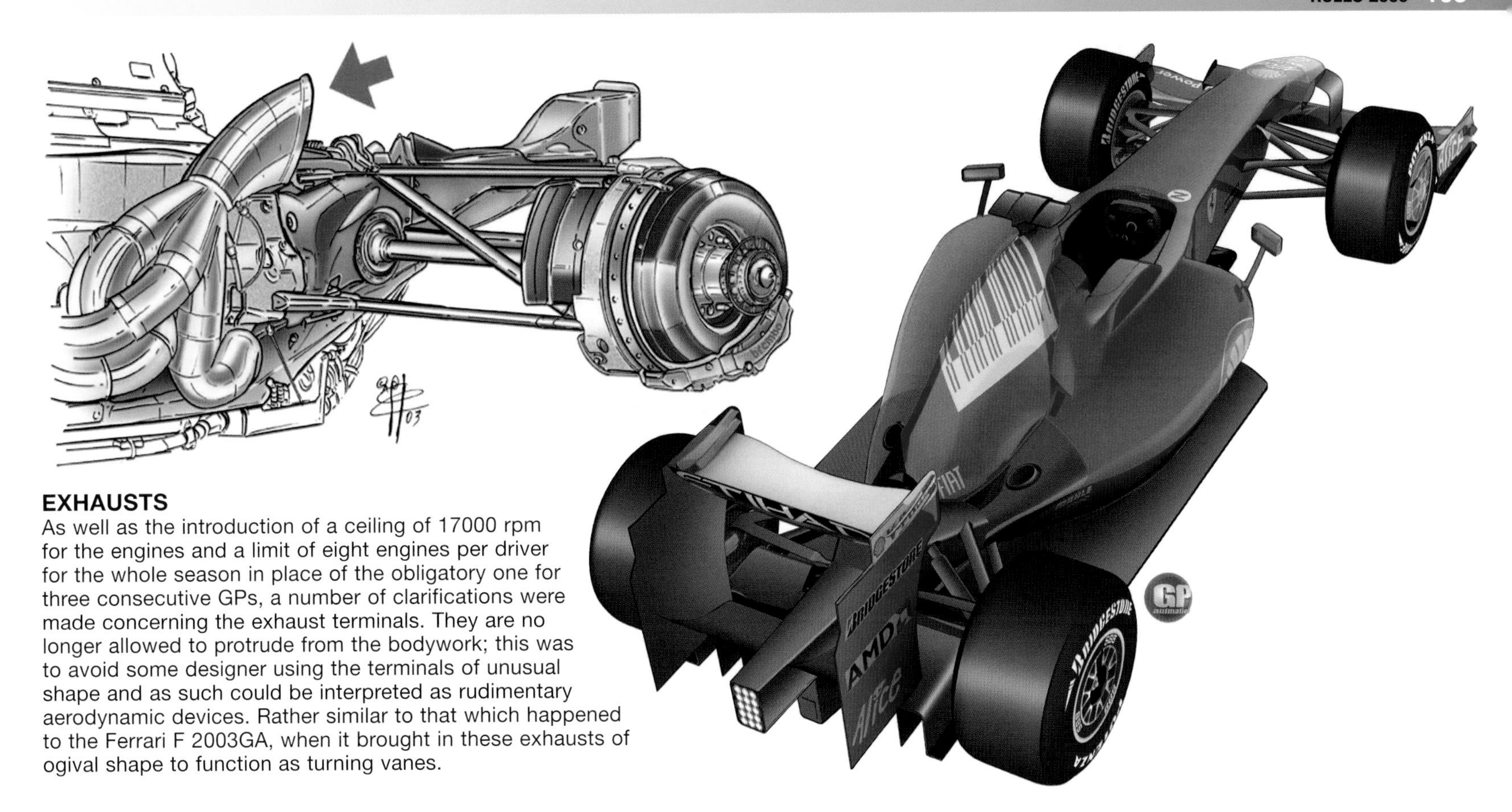

EXHAUSTS

As well as the introduction of a ceiling of 17000 rpm for the engines and a limit of eight engines per driver for the whole season in place of the obligatory one for three consecutive GPs, a number of clarifications were made concerning the exhaust terminals. They are no longer allowed to protrude from the bodywork; this was to avoid some designer using the terminals of unusual shape and as such could be interpreted as rudimentary aerodynamic devices. Rather similar to that which happened to the Ferrari F 2003GA, when it brought in these exhausts of ogival shape to function as turning vanes.

KERS

Another new feature is the introduction of the non-obligatory KERS (kinetic energy recovery system) to make overtaking easier. In practice, drivers can accumulate energy under braking for a total of 60 kwj, equal to 80 hp to then discharge it for a maximum of not more than 6.6 seconds a lap. There are two systems of energy recovery: one is mechanical and is used only by Williams (a flywheel placed beside the gearbox) and the other is electric, a system used by all the other teams. The illustration shows the two distinct phases that should enable the drivers to use an overboost to get past other cars more easily.
In the braking phase (1) the electric motor (2) connected to a drive shaft (so increasing the engine brake effect) acts as a generator and progressively charges the lithium batteries (4), all of which is controlled by an electronic management system (3). Above are the stages of energy accumulation, below the overboost phase. The driver can manually manage the energy accumulated by the batteries with the electronic management system, which commutes the method of operation from generator to electric motor, discharging an added power of about 80 hp for the 6.6 second maximum. With a button (5) on the steering wheel, the driver can decide how to use it depending on necessity; whether to use all the energy at once or at various points on the circuit, but always for that maximum total of 6.6 seconds a lap.

The technical revolution imposed by the Federation for the 2009 season centred on the objective of prodding the imagination of the designers to the point that never have such varied cars been presented, so much so as to make it extremely easy to recognise the different racers.
If on the one hand there were major new developments with a view to moving the pre-established hierarchy of recent seasons, on the other the Federation completely failed on the cost reduction front and the drastic aerodynamic limitation of the cars.
The 2009 season will pass into history as the most expensive of all those that have gone before, both for the development of KERS (over 50 million euros) and the excessive aerodynamic research necessary to recover the downforce that had to be drastically reduced.
At the introduction of the new regulations they were much hoping that the 2009 cars would be almost totally without aerodynamic appendages and with more simplified front wings and diffusers, as was previewed with the designs in the winter break.
Looking at the cars of the 2009 season, it can easily be seen that this spirit was in no way respected if not for the absence of the chimneys, flaps and louvers in the upper part of the body.
The front wings have become even more complex than those seen until 2008: the turning vanes inside the suspension have made a comeback, but located under the nose, the mirrors acted as phoney pretexts for the return of aerodynamic devices (see Ferrari, BMW, Brawn and Force India).
As early as just before Christmas it was clear that the text of the regulations had left many "grey" zones which were not well defined that could have given rise to interpretations in such a way as to distort the spirit of the regulations themselves.
On the other hand, the task of every team is to carefully study the regulations in an attempt to find loopholes for new solutions.
As often happens in the changes of regulations, there were those who pushed the rules to the limit, and that appeared evident right from the first tests of the new cars on the track.
It was from mid-January that disgruntlement began to spread due to the interpretation given to the realisation of the Toyota diffusers, with the controversy leading inevitably to the first race of the season in Australia and ending up in a Paris appeals court.
The first warning came with the Ferrari exhausts episode, which immediately died a death because it concerned a clarification of the original text by which Maranello had been inspired. At the subsequent launches, it was noticed that another two teams, Williams and Brawn, had interpreted the diffuser regulations in a much different way.

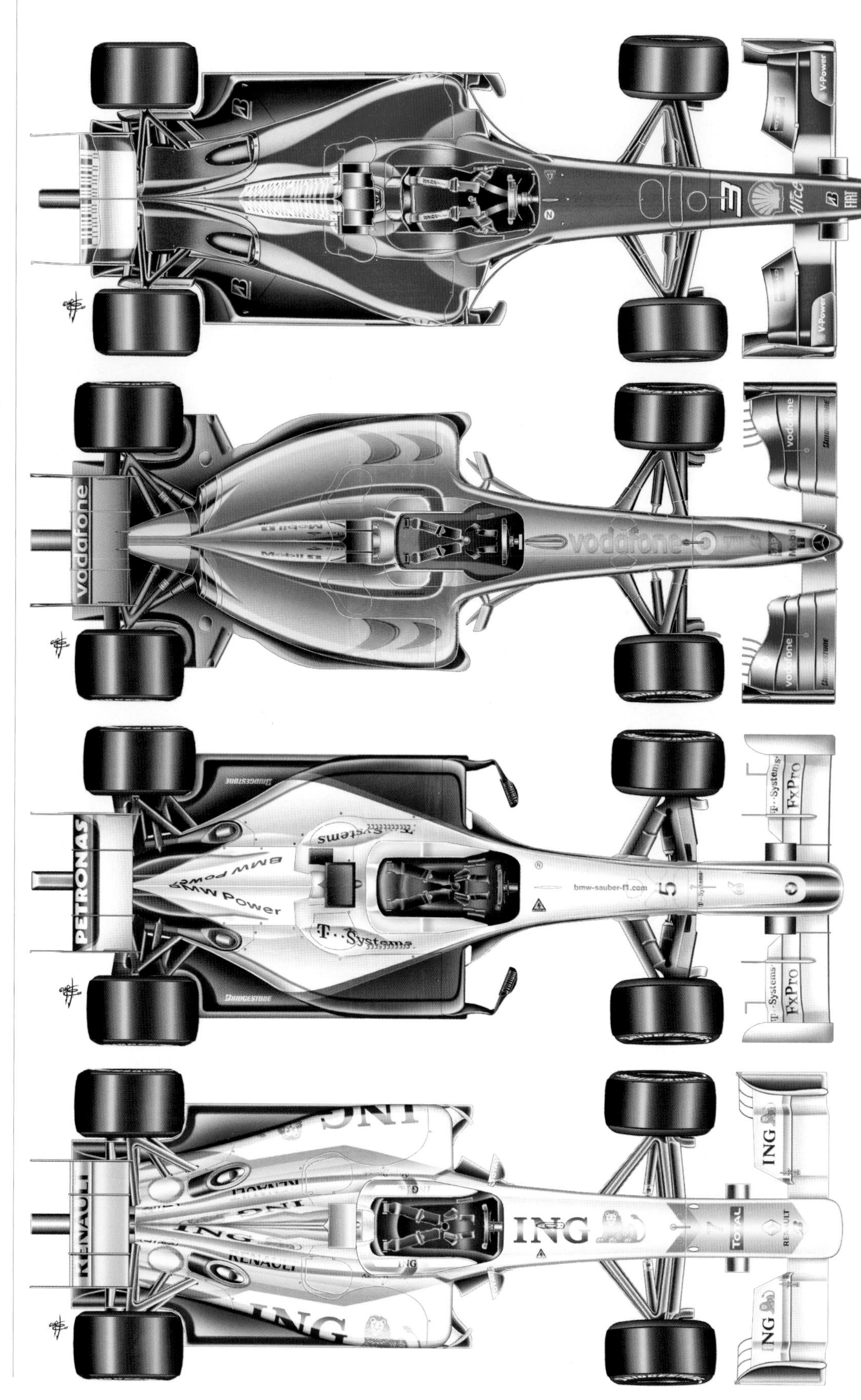

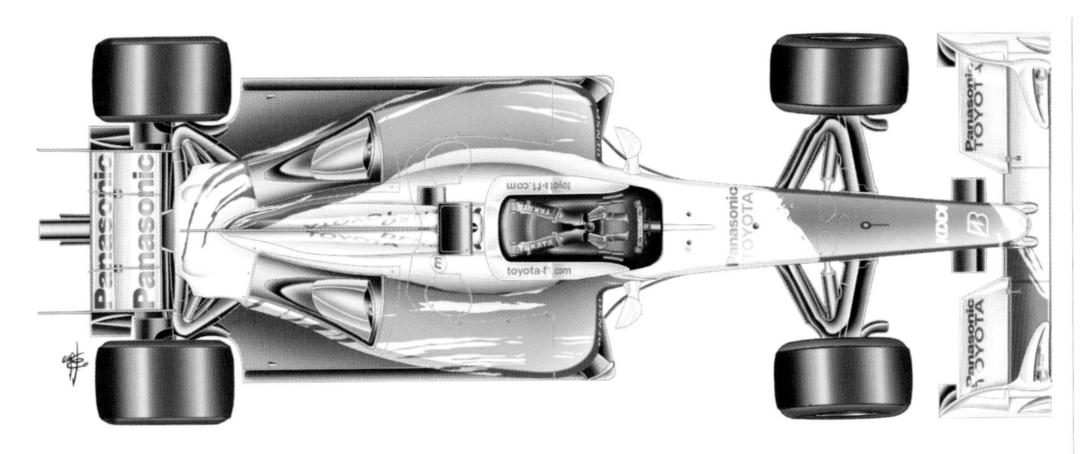

Strangely, Charlie Whiting has given his approval to the contrivance produced by the three teams, while all the others had respected the text to the letter.
Despite his fame as a transgressive designer by taking development to the limit, Adrian Newey presented the latest Red Bull with diffusers perfectly in line with the rules, as had Ferrari, BMW, McLaren and Renault.
The new regulations have many areas that permitted the return of aerodynamic solutions which, at first, seemed destined to disappear. That was the case with the raised mini-flap applied to the front wing end plates, the turning vanes in the lower area of the purposely raised nose, of a large turning vane in the zone in front of the sidepods (where in theory the regulations provided for a much smaller one) and above all two vertical turning vanes to support the mirrors positioned to overhang the 'pods.
The constraints around the front end concern the area above the front wing and the sides of the chassis, but the legislators had not thought of the lower zone. This is where there was the regular application of aerodynamic appendages on all the new cars that had raised the chassis a lot and, therefore, the room to host them.
More astute was the idea of bigger turning vanes in relation to those hypothesised in front of the sidepods and of the vertical ones at their sides.
In delineating the prohibited perimeter, the Federation took as reference the dimensions of the 2008 sidepods but, for example in Ferrari's case, Maranello opted for a shortening of the sidepods of their 2009 car to distance turbulence caused by the front wheels.
This important decision freed space that permitted the lengthening of the turning vane in the area close to the monocoque and, even more important, the vertical ones supporting the mirrors.
The latter, positioned outside the sidepods themselves needed anchorage in the lower area, which in turn exploited another norm saying it must exist because every element that protrudes in the upper zone of the car must have a shadow plate in the lower area. These vertical fins also already seen from the beginning on the Toyota and Williams in Melbourne, proliferated during the season.

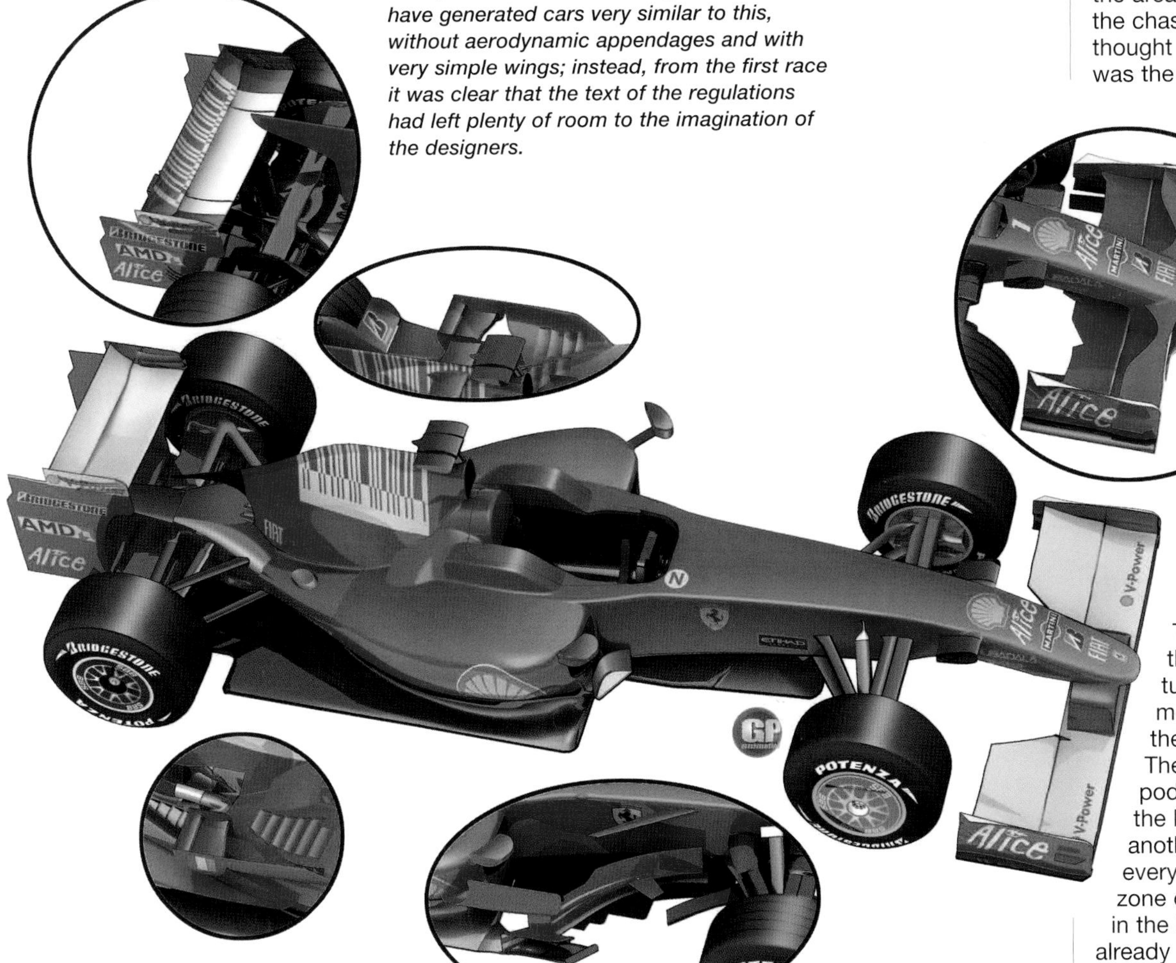

The spirit of the new 2009 regulations should have generated cars very similar to this, without aerodynamic appendages and with very simple wings; instead, from the first race it was clear that the text of the regulations had left plenty of room to the imagination of the designers.

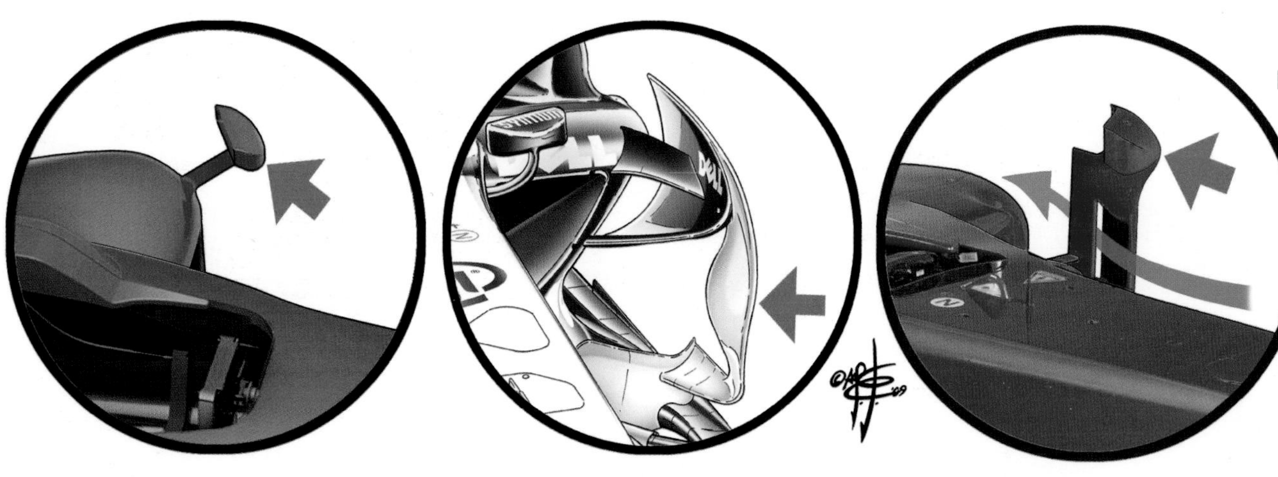

MIRRORS

These had become an excuse for the return of the vertical aerodynamic appendages through the back door, as it were. They were introduced by BMW in 2008 and are, obviously, generalised features after Ferrari became the first to adopt them.

FERRARI

The lines of the Ferrari F60 were extremely clean, especially when compared with the F2008.
At the car's launch and first outings it had exhausts that protruded from the body. They were later banned after a clarification by the Federation (see the 2009 Regulations chapter).

Two teams immediately stepped into the spotlight with new cars that were impossible to copy, having introduced concepts that require a new global design of the whole car. They were Adrian Newey's Red Bull and Brawn, the circus's new arrival under the knowledgeable guidance of Ross Brawn. The former extreme, in full respect of the traditions of the British genius, could be considered the most innovative and radical of the circus, even if two of the features are nothing more than the return of solutions already seen but by now already forgotten by F1. One thinks in particular of the rear pull rod layout suspension of the rear axle and the V-shape of the chassis in its front section. With new constraints imposed by the Federation on the diffusers, it became very important to reduce the bulk in the upper part of the terminal area of the car and less so in the lower zone. That is why, with the linked layout and all the elements down low, the gearbox-differential group was able to be lowered by about 15 cm (an enormous amount), and even the upper wishbone of the rear suspension transformed itself into a further aerodynamic element in synergy with the diffuser and the lower plane of the rear wing. As far as the section of the chassis is concerned, the regu-

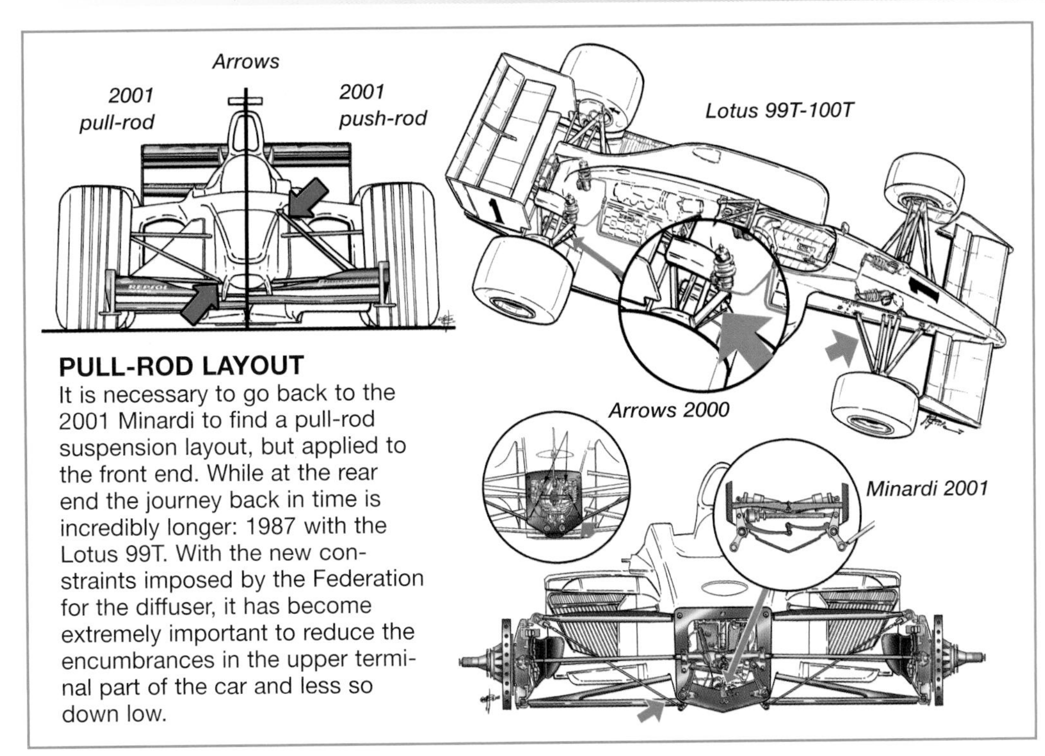

PULL-ROD LAYOUT

It is necessary to go back to the 2001 Minardi to find a pull-rod suspension layout, but applied to the front end. While at the rear end the journey back in time is incredibly longer: 1987 with the Lotus 99T. With the new constraints imposed by the Federation for the diffuser, it has become extremely important to reduce the encumbrances in the upper terminal part of the car and less so down low.

McLAREN

The flat bottom cut in front of the McLaren's rear wheels was new and was first seen on the M24's second track appearance. The car immediately showed that was wrong and that forced a great search for improvement, with many new developments at every race.

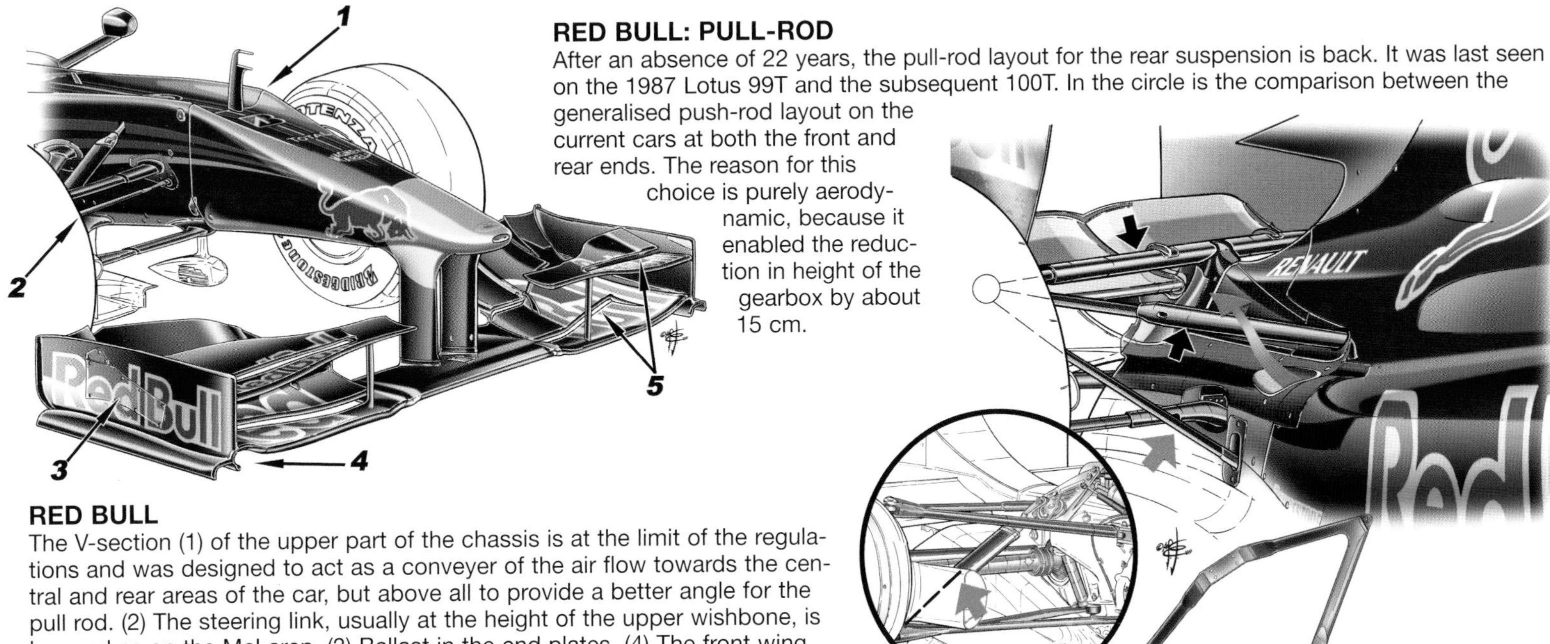

RED BULL: PULL-ROD

After an absence of 22 years, the pull-rod layout for the rear suspension is back. It was last seen on the 1987 Lotus 99T and the subsequent 100T. In the circle is the comparison between the generalised push-rod layout on the current cars at both the front and rear ends. The reason for this choice is purely aerodynamic, because it enabled the reduction in height of the gearbox by about 15 cm.

RED BULL

The V-section (1) of the upper part of the chassis is at the limit of the regulations and was designed to act as a conveyer of the air flow towards the central and rear areas of the car, but above all to provide a better angle for the pull rod. (2) The steering link, usually at the height of the upper wishbone, is lowered as on the McLaren. (3) Ballast in the end plates. (4) The front wing group is very sophisticated (5) right from its debut. Newey has adopted the raised flaps with large blowers to increase efficiency.

lations impose minimum sizes. Newey played around with a reduction at the centre and raising the sides to maintain the same total section but gaining considerably in aerodynamic penetration and installation of the suspension. Almost all the 2009 cars have the nose raised from the ground (the Red Bull drivers have their feet about 3 centimetres higher) with the angles of the suspension wishbones less favourable. This Newey solution enables the suspension pull rod to be positioned up higher and maintain the same angulation as that of 2008. Instead, the Brawn moves by opposing concepts: no exaggerations, but a knowledgeable mix of concrete solutions. Its nose is the lowest of all the 2009 cars, to the advantage of a much reduced centre of gravity and an optimum suspension geometry. The wheelbase is reduced, as is the bulk of the sidepods (putting all their money on the absence of KERS) and a weight distribution more easily manageable in line with needs are the other peculiarities of this extraordinary car.

To all of this we add many intelligent features in aerodynamics, like the sophisticated design of the front wing end plates, the divergence in the lower part of the chassis and especially the two-level diffuser (under observation at Melbourne).

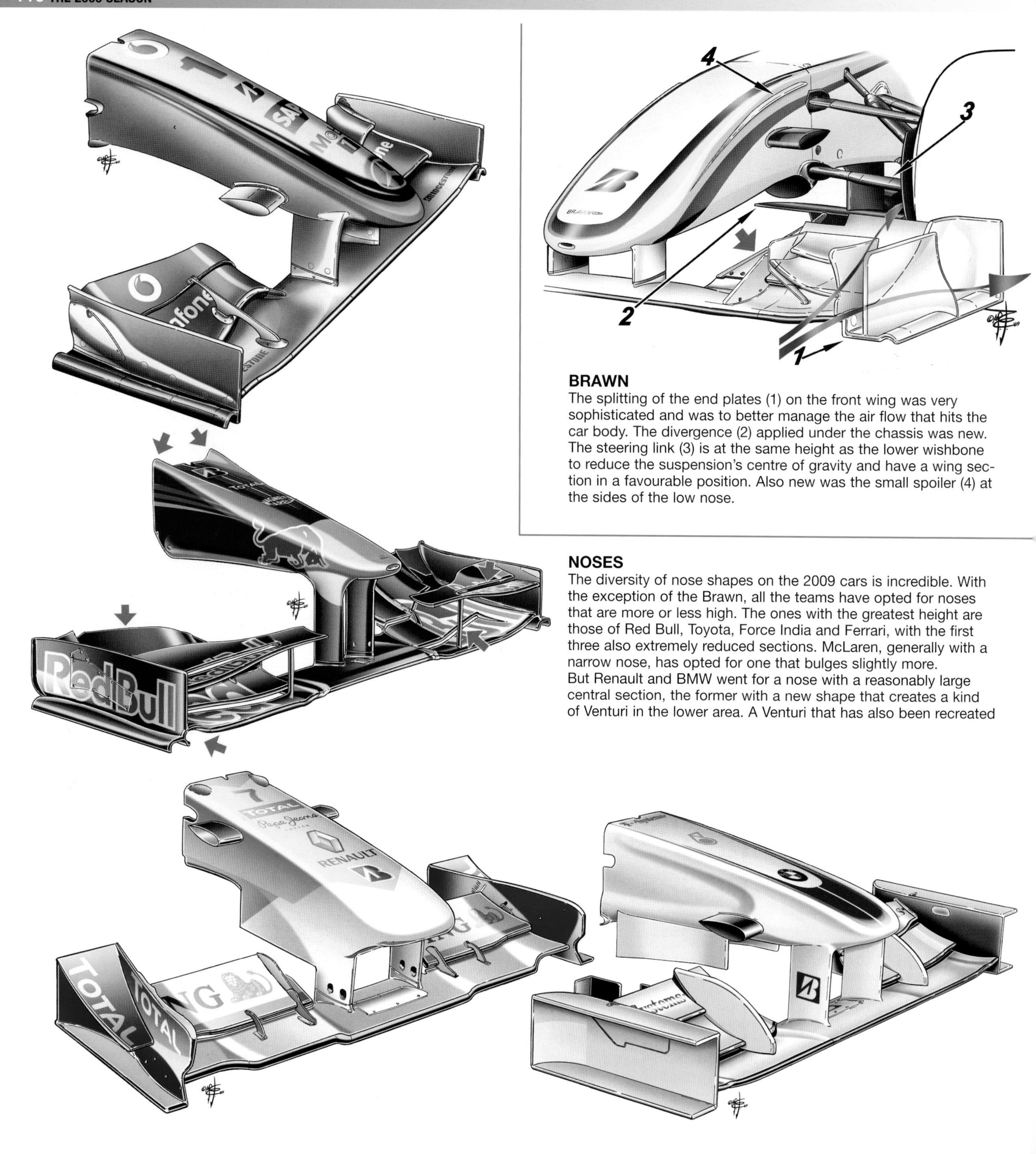

BRAWN
The splitting of the end plates (1) on the front wing was very sophisticated and was to better manage the air flow that hits the car body. The divergence (2) applied under the chassis was new. The steering link (3) is at the same height as the lower wishbone to reduce the suspension's centre of gravity and have a wing section in a favourable position. Also new was the small spoiler (4) at the sides of the low nose.

NOSES
The diversity of nose shapes on the 2009 cars is incredible. With the exception of the Brawn, all the teams have opted for noses that are more or less high. The ones with the greatest height are those of Red Bull, Toyota, Force India and Ferrari, with the first three also extremely reduced sections. McLaren, generally with a narrow nose, has opted for one that bulges slightly more.
But Renault and BMW went for a nose with a reasonably large central section, the former with a new shape that creates a kind of Venturi in the lower area. A Venturi that has also been recreated

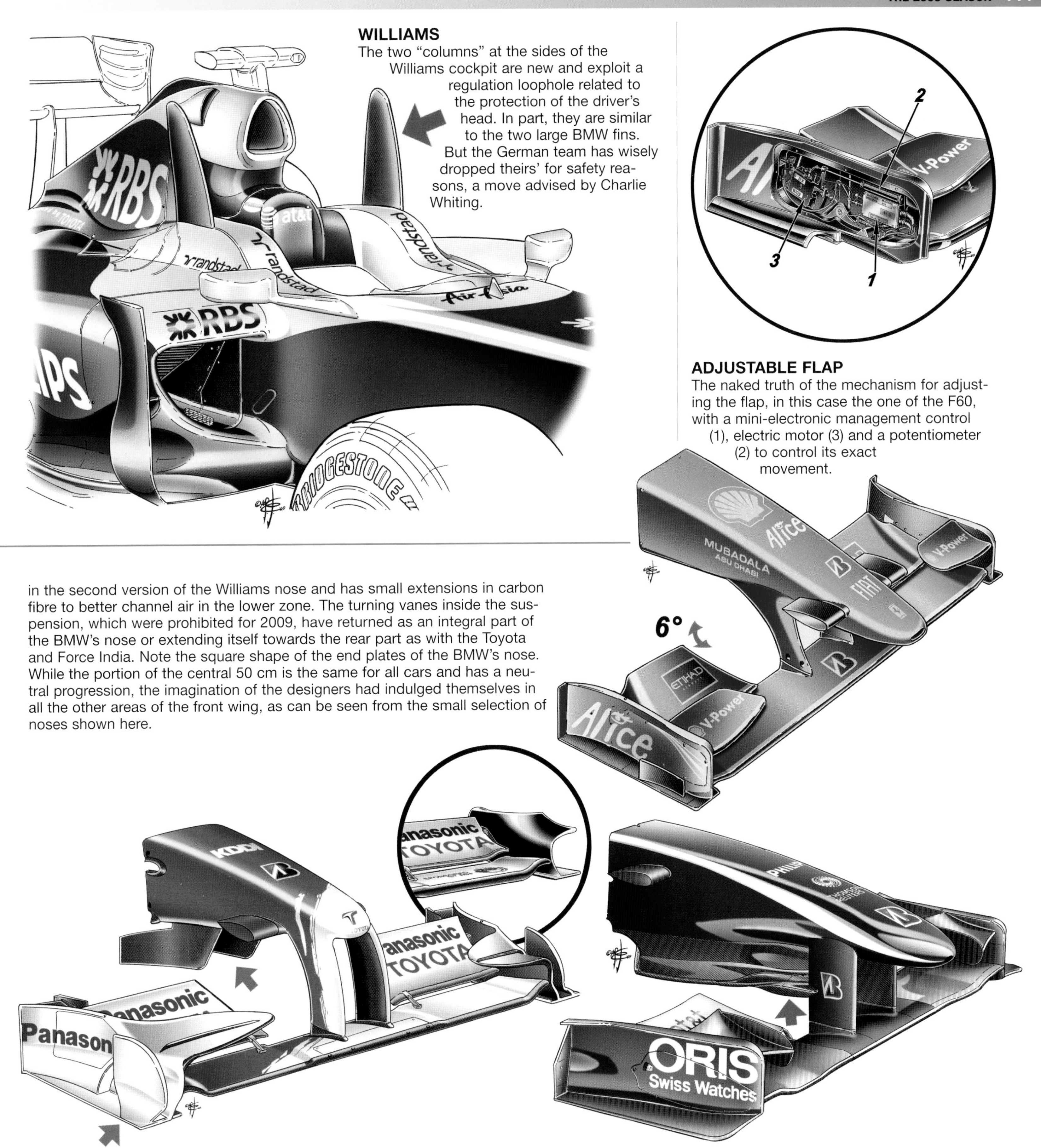

WILLIAMS

The two "columns" at the sides of the Williams cockpit are new and exploit a regulation loophole related to the protection of the driver's head. In part, they are similar to the two large BMW fins. But the German team has wisely dropped theirs' for safety reasons, a move advised by Charlie Whiting.

ADJUSTABLE FLAP

The naked truth of the mechanism for adjusting the flap, in this case the one of the F60, with a mini-electronic management control (1), electric motor (3) and a potentiometer (2) to control its exact movement.

in the second version of the Williams nose and has small extensions in carbon fibre to better channel air in the lower zone. The turning vanes inside the suspension, which were prohibited for 2009, have returned as an integral part of the BMW's nose or extending itself towards the rear part as with the Toyota and Force India. Note the square shape of the end plates of the BMW's nose. While the portion of the central 50 cm is the same for all cars and has a neutral progression, the imagination of the designers had indulged themselves in all the other areas of the front wing, as can be seen from the small selection of noses shown here.

THE CASE OF THE DIFFUSERS

The diffuser controversy blew up as expected in Melbourne for the three teams which were called “the hole gang”: Brawn, Williams and Toyota. The spirit of the regulation was to severely limit the diffusers, imposing a single height and length for its entire width.
Seven teams respected the technical text to the letter, as can be seen in the example of the lower part of the Ferrari F60. The other three, starting with Toyota, then Williams and Brawn made their debut with a double central channel inspired by that of the car which, until the previous season, was fed by an ample window and was rather developed in height. But if the loophole in the regulations to make the deformable structure pass in the central 15 cm (Toyota) permits a mini-central channel to be placed (always in disaccord with the spirit of the regulation) the question is more complicated related to the hole that these three teams had to make to feed the new double central channel. The regulation, in articles from 3.121 to 7, is extremely complex and contorted. A hole is permitted in the vertical plane that joins the two horizontal planes formed by the reference plane, provided no element of the bodywork can be seen looking at the car vertically from underneath.
All that is needed is a gap of one single millimetre and that makes it clear how the wording of the regulation splits hairs. A hole in the horizontal plane is also tolerated in the stepped section provided it is an element of the suspension which, at that point, acts as a vertical part of the same. For this, both Williams and Brawn have fairly short bodies, while Toyota has a longer one.
The case ended up in the appeals court of Paris and that legitimated the hole solution on the eve of the third race of the season, forcing all the other teams to follow.

RACE TO LEARN
.ORG

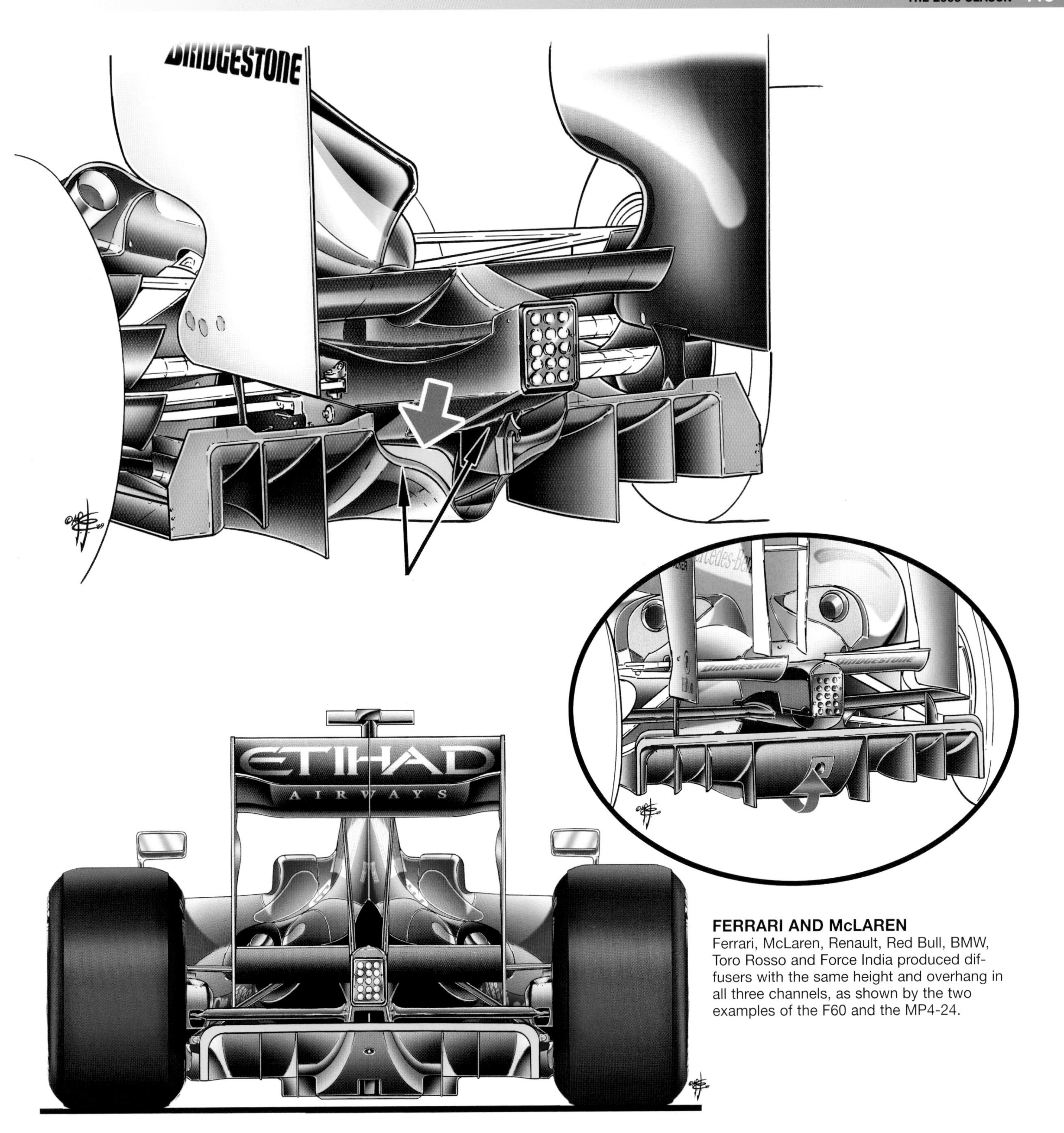

FERRARI AND McLAREN
Ferrari, McLaren, Renault, Red Bull, BMW, Toro Rosso and Force India produced diffusers with the same height and overhang in all three channels, as shown by the two examples of the F60 and the MP4-24.

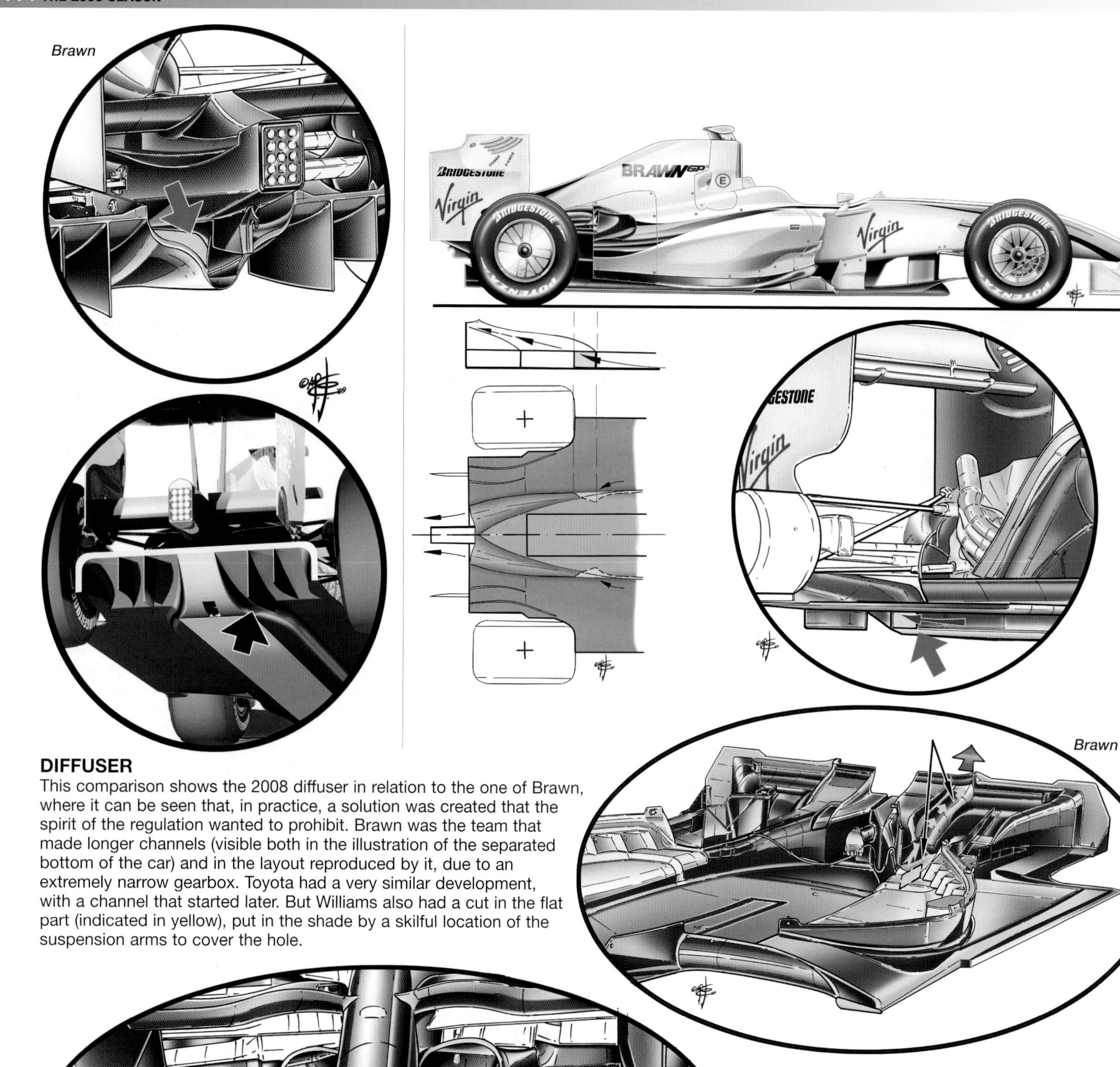

DIFFUSER

This comparison shows the 2008 diffuser in relation to the one of Brawn, where it can be seen that, in practice, a solution was created that the spirit of the regulation wanted to prohibit. Brawn was the team that made longer channels (visible both in the illustration of the separated bottom of the car) and in the layout reproduced by it, due to an extremely narrow gearbox. Toyota had a very similar development, with a channel that started later. But Williams also had a cut in the flat part (indicated in yellow), put in the shade by a skilful location of the suspension arms to cover the hole.

DIFFUSERS

In practice, they had gone back to a blown central channel that was not in line with the two laterals, as happened in a more evident way until 2008 (the central channel was longer and much higher). The example is that of McLaren (see the 2007 Technical Analysis), which partially cut the lateral channels (1), exploiting the faired suspension arm (2) as a superior limit.

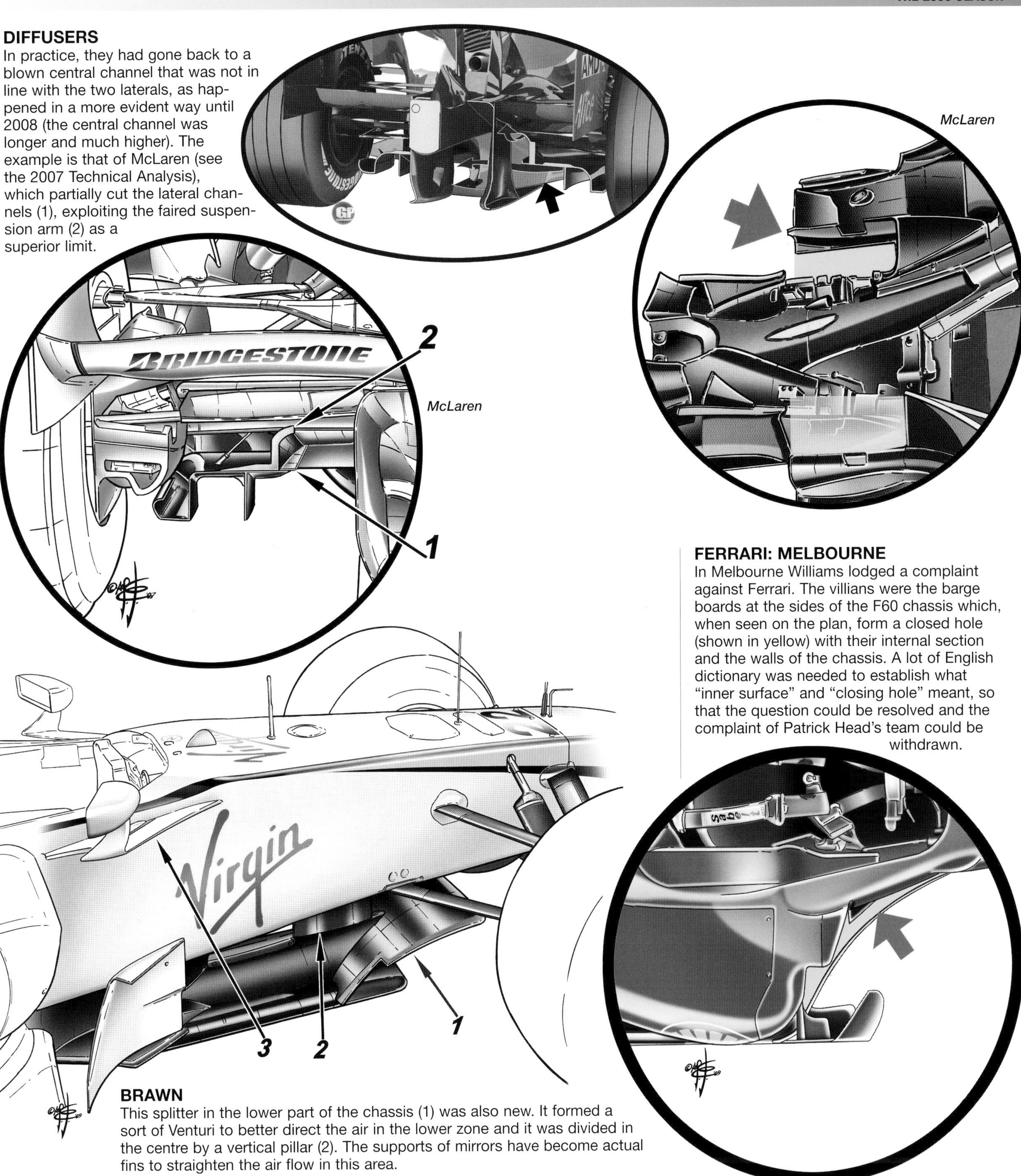

FERRARI: MELBOURNE

In Melbourne Williams lodged a complaint against Ferrari. The villians were the barge boards at the sides of the F60 chassis which, when seen on the plan, form a closed hole (shown in yellow) with their internal section and the walls of the chassis. A lot of English dictionary was needed to establish what "inner surface" and "closing hole" meant, so that the question could be resolved and the complaint of Patrick Head's team could be withdrawn.

BRAWN

This splitter in the lower part of the chassis (1) was also new. It formed a sort of Venturi to better direct the air in the lower zone and it was divided in the centre by a vertical pillar (2). The supports of mirrors have become actual fins to straighten the air flow in this area.

McLAREN

These brake air ducts introduced by McLaren led the way in the 2008 pre-season tests but were never used afterwards. The regulation in this sector became more permissive and many teams used them, even if in slightly difference shapes.

RENAULT: SHANGAI

The new Renault bottom with double the diffusers in the central area arrived in China just in time on the Friday and was fitted to a new body that required additional insulation panels to ensure it did not go up in flames, as happened during Saturday morning testing. Note the width of the central channel that exploits the maximum dimensions permitted by the regulations of 50 cm.

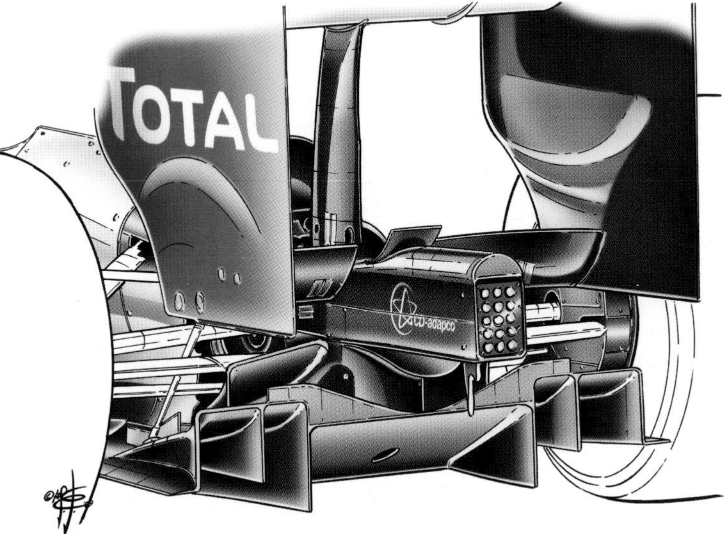

FERRARI: KERS

The KERS short circuit that happened to Kimi Raikkonen in Friday morning practice in Malaysia allowed observers to see the various components of the system and, above all, the dimensions and the disposition of the battery group. The Lithium battery pack is contained in a voluminous covering of carbon fibre that guarantees good insulation. To facilitate its substitution, the chassis has a sort of large niche in the lower area, which moves the location of the fuel tank upwards by at least 12-15 cm and, as a result, its centre of gravity. This structure weighs over 25 kg.

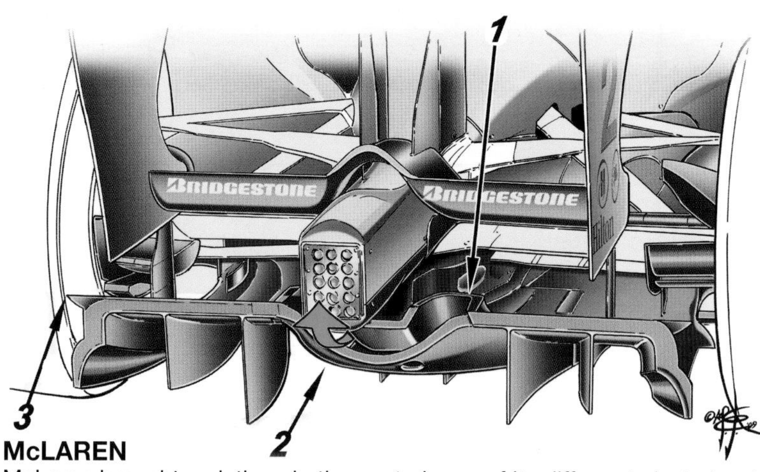

McLAREN

McLaren brought variations in the central area of its diffuser to just about every race. The one we show was introduced in Spain and was the fourth evolution of the diffuser by the team, starting from the basics of the one that came out in Australia, of which it retained the shape of the lower area (2). The section of the double central channel is not very large (2). The small finlet (3) at the sides of the central channel was also new and recouped a little downforce in the zone near the wheels.

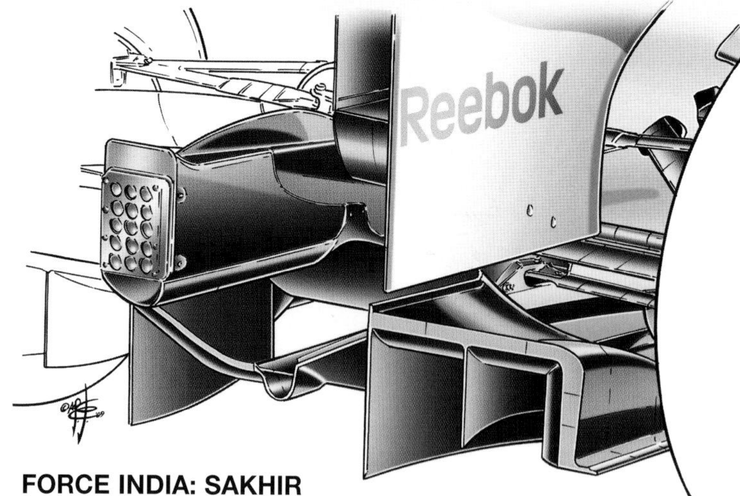

FORCE INDIA: SAKHIR

Force India was the second team after Renault to join the "hole gang" with a very similar solution to the one brought in by the French team in China. The central channel exploits the 50 cm maximum width permitted by the regulations : the new aerodynamic package included a revised detail of the front wing in the low area of the sidepods.

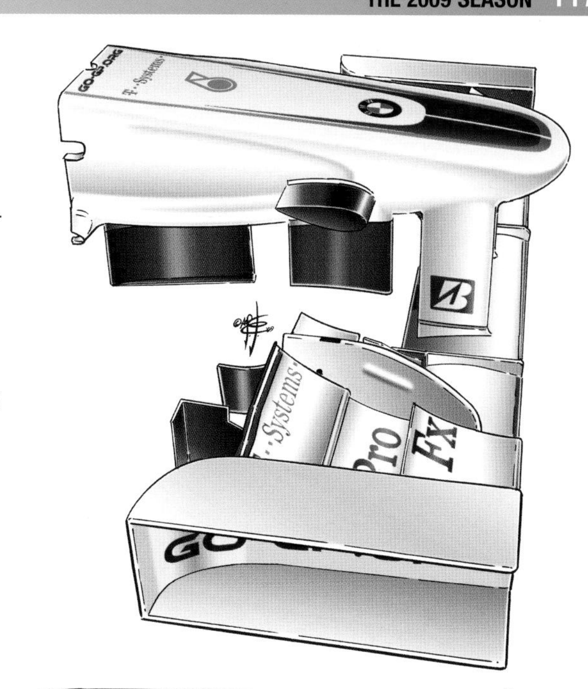

BMW: BARCELONA

BMW fielded a practically new version of their car without KERS in Spain. The nose was redesigned and underwent considerable crash testing. It now points much higher and has four vertical end plates in the lower area. The sidepods had been drastically slimmed down below, eliminating the KERS batteries. In comparison, the new sidepods were much more tapered (1), as on last year's car. The old 'pods had a more vertical opening so that the KERS batteries (2) could be accommodated at the sides; the vertical deflector (3) was also new, as was the plane (4) at mid-height, to recover downforce.

FERRARI: BARCELONA

Ferrari also had an almost B version of their car in Spain on which a considerable amount of work had been carried out in the lower area of the gearbox and the hydraulic system. Indispensable was the work to create the new two-stage diffuser in the central zone, inspired by that of the Brawn. It is fed by an extremely advanced hole (1) and has a highly considerable section (2) but did not have a raised area (3) in its trailing edge. The intermediate end plates had disappeared from the lateral channels.

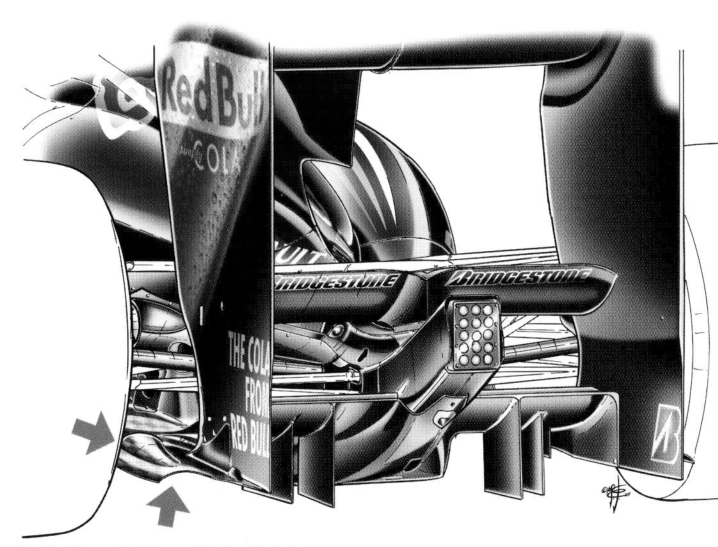

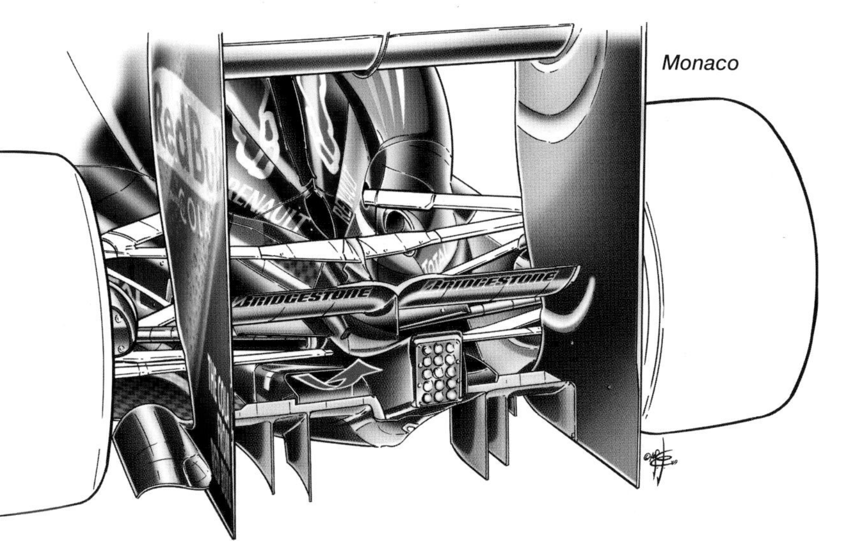

RED BULL: DIFFUSERS

Red Bull was the last among the top teams to introduce the double central diffuser. It did so at Monaco without distorting the layout of the extremely low rear end (the old one is on the left). It has a very similar layout to the one brought in by Ferrari and does not protrude in either height or length in relation to the other channels. The curved mini-channel (indicated by the arrows) was retained in the area close to the wheels.

RED BULL RB5 "B": SILVERSTONE

The season's turning point came at Silverstone with the debut of the "B" version of the Red Bull, very different to the intermediate cars that had appeared at Monaco. The narrow nose was abandoned in favour of one designed to improve air flow to the underside of the car. The exaggeration of the flat configuration of the nose meant that the video cameras were integrated into the structure and acted as the first aerodynamic profile channelling the air flow in this area. With no realistic opportunity to revise the gearbox and the rear suspension, the team opted to distance these elements from the delicate diffuser area, by slightly shifting (1) the rear axle (as can be seen from the view from below); the same thing was done at the front in order to maintain the same wheelbase. This operation gained a small but precious amount of space in which a more aggressive lower aerodynamic package could be installed, along with larger intakes (2) for the central channels. In place of the dual central diffuser there was now a three-channel element similar to that of the Toyota. The rear wing end-plates were also new with sculpted lower sections, but were still connected to the diffuser's lateral channels.

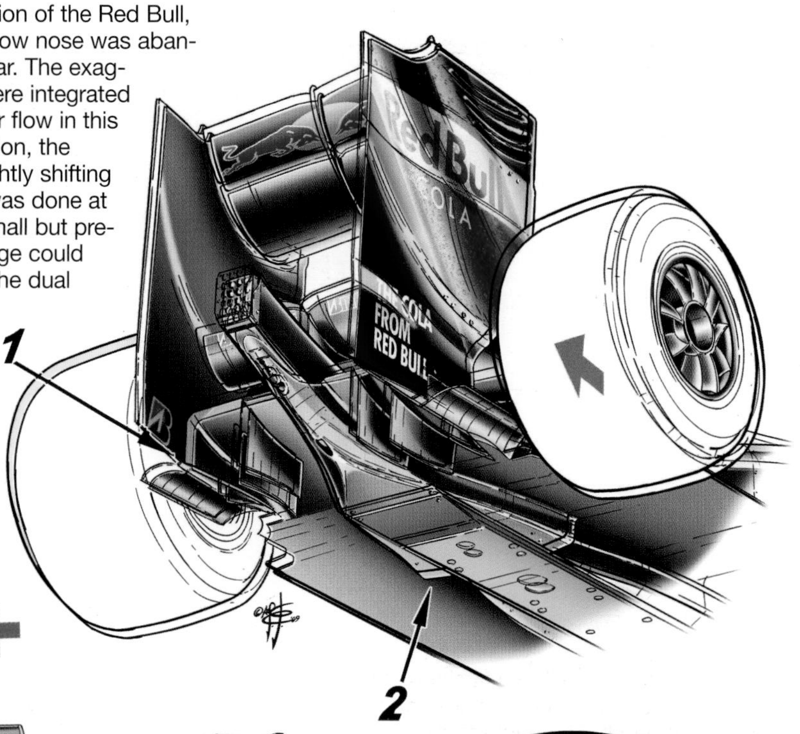

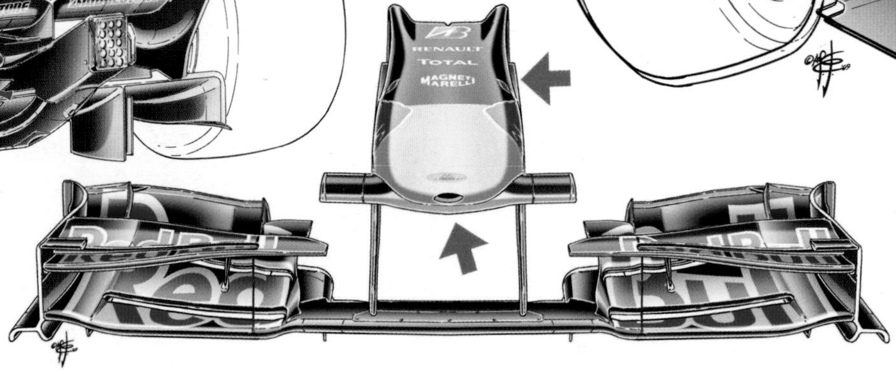

TOYOTA: ISTANBUL

In Turkey, Toyota radically modified the TF109's diffuser, abolishing the mini-channel (within the 15 cm exempt from the restrictions) (1), completely revising the dual central channel with trailing edge (2) further forwards than that of the original design. The real surprise came with these vertical slots (3) interfering with the turbulence generated by the rear wheels.

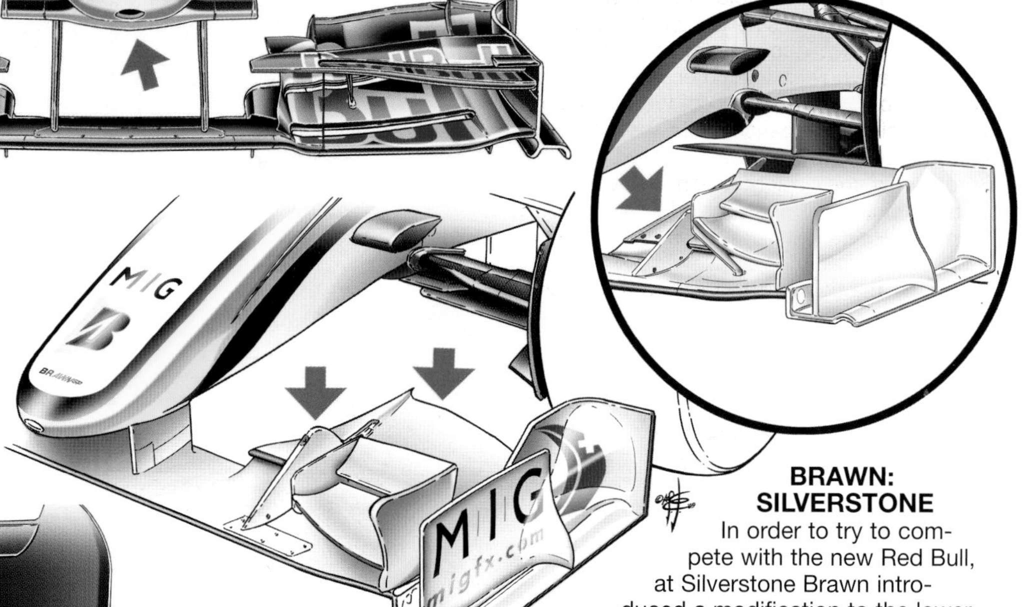

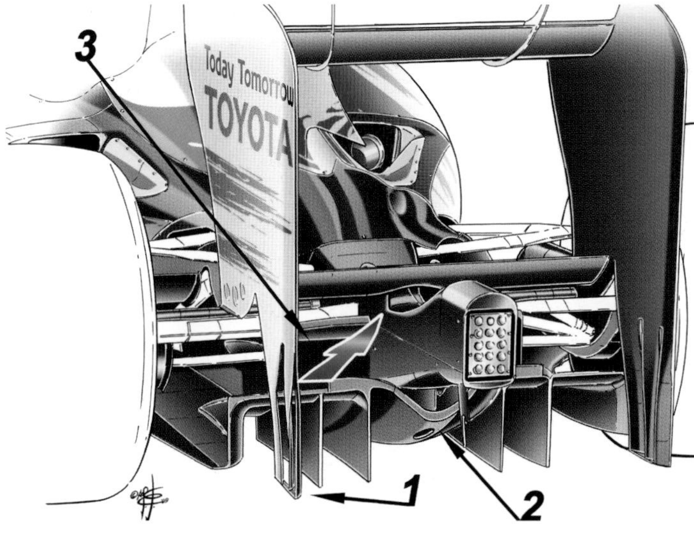

BRAWN: SILVERSTONE

In order to try to compete with the new Red Bull, at Silverstone Brawn introduced a modification to the lower part of the end-plates with a new front wing (already modified for the previous race in Turkey with a single flap), characterised by a square-cut section and two lower slides. All this was designed to provide as much front end downforce as possible.

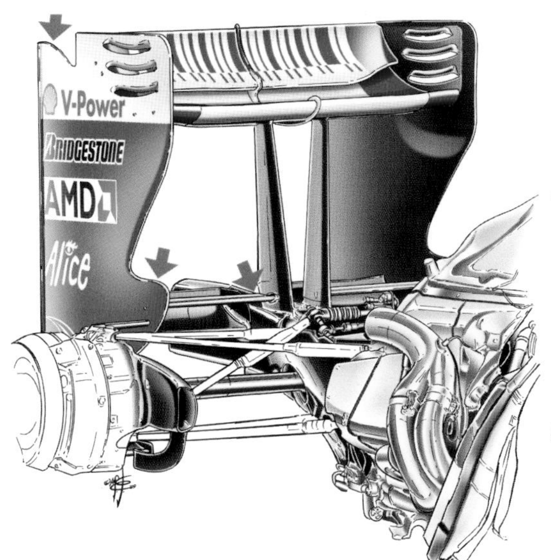

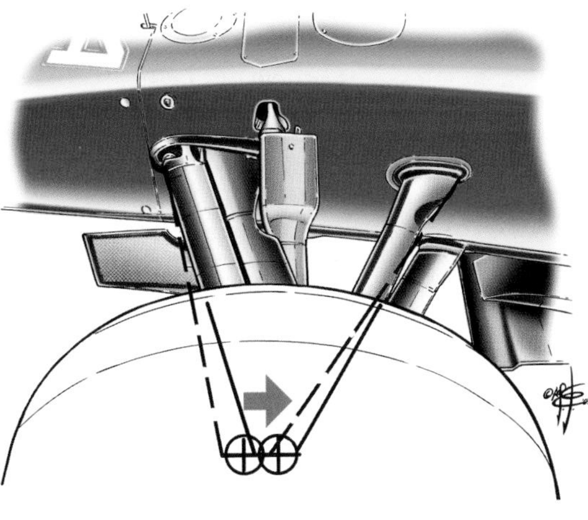

FERRARI: SILVERSTONE
Again at Silverstone, Ferrari presented a significant development after having introduced a new rear wing in Istanbul characterised by end-plates with a lower extension and a cut in the upper rear part. The team in fact shortened the wheelbase to shift weight onto the front axle, angling the suspension arms backwards. The nose and the turning vanes in front of the sidepods were new. At the next race at the Nürburgring, an extractor profile was introduced characterised by three intermediate vanes (right) rather than the previous element (left).

McLAREN: NÜRBURGRING
At the Nürburgring, a single radically modified MP4-24 was entered for Hamilton: a true "B" version. A correction was made to the front wing that had been unable to expel much air towards the area outside the tyres. Above all, the whole central and rear sections of the car were revised. Lower, straighter sidepods, thanks in part to a lowering of the exhausts (in the circle a comparison with the old, more sinuous configuration).
Surprisingly, the designers eliminated the unique cut in the underside in the area ahead of the sidepods where the tapering Coke-bottle effect was reduced to allow for an alignment with the diffuser's wider central channel.

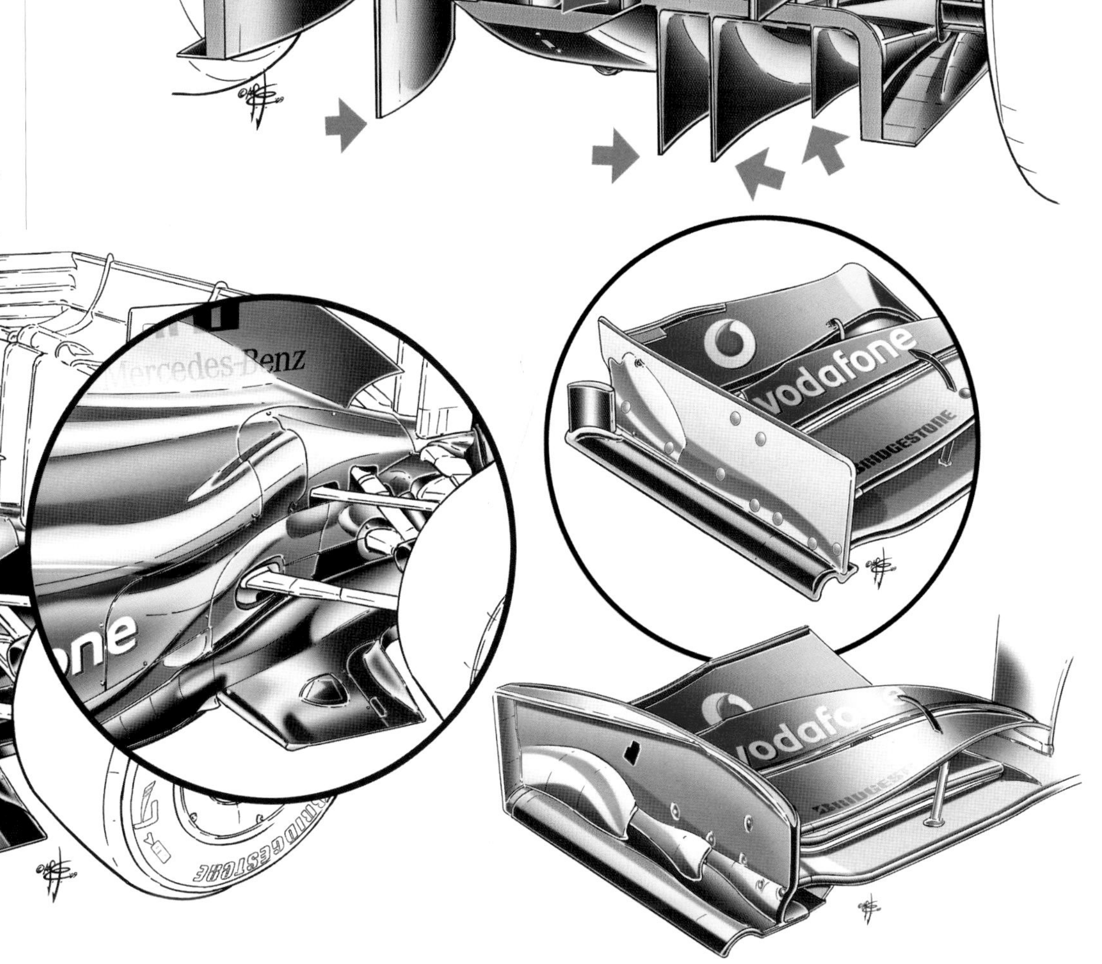

Giorgio Nada Editore Srl

Editorial manager
Leonardo Acerbi

Editorial coordination
Diana Calarco

Product development
Studio Enigma

Graphic design and cover
Aimone Bolliger

Contributors
Franco Nugnes (engines)
Michele Merlino (engine tables)
Kazuiko Kasai and Mark Hughes (tyres)

Computer graphic
Belinda Lucidi
Alessia Bardino
Cristina Ravetta
Marco Verna
Paolo Rondelli

3D animations
Generoso Annunziata

Printed in Italy by
Grafiche Flaminia Srl
Foligno (PG)
september 2009

Giorgio Nada Editore
Via Claudio Treves,15/17
I - 20090 VIMODRONE MI
Tel. +39 02 27301126
Fax +39 02 27301454
e-mail: info@giorgionadaeditore.it
www.giorgionadaeditore.it

Distribution
Giunti Editore SpA
via Bolognese 165
I - 50139 FIRENZE
www.giunti.it

Formula 1 2008/2009 - technical analysis
ISBN: 978-88-7911-466-0